ACCOUNTING FOR HEALTH AND HEALTH CARE

APPROACHES TO MEASURING THE SOURCES AND COSTS OF THEIR IMPROVEMENT

Panel to Advance a Research Program on the Design of National Health Accounts

Committee on National Statistics

Division of Behavioral and Social Sciences and Education

NATIONAL RESEARCH COUNCIL
OF THE NATIONAL ACADEMIES

THE NATIONAL ACADEMIES PRESS
Washington, D.C.
www.nap.edu

THE NATIONAL ACADEMIES PRESS 500 Fifth Street, N.W. Washington, DC 20001

NOTICE: The project that is the subject of this report was approved by the Governing Board of the National Research Council, whose members are drawn from the councils of the National Academy of Sciences, the National Academy of Engineering, and the Institute of Medicine. The members of the committee responsible for the report were chosen for their special competences and with regard for appropriate balance.

This study was supported by contract number N01-OD-4-2139 between the National Academy of Sciences and the U.S. Department of Health and Human Services. Support for the work of the Committee on National Statistics is provided by a consortium of federal agencies through a grant from the National Science Foundation (award number SES-0453930). Any opinions, findings, conclusions, or recommendations expressed in this publication are those of the author(s) and do not necessarily reflect the view of the organizations or agencies that provided support for this project.

Library of Congress Cataloging-in-Publication Data

National Research Council (U.S.). Panel to Advance a Research Program on the Design of National Health Accounts.
Accounting for health and health care : approaches to measuring the sources and costs of their improvement / Panel to Advance a Research Program on the Design of National Health Accounts, Committee on National Statistics, Division of Behavioral and Social Sciences and Education, National Research Council.
p. ; cm.
Includes bibliographical references.
ISBN 978-0-309-15679-0 (pbk.) — ISBN 978-0-309-15680-6 (pdf) 1. Medical care, Cost of—United States. I. Title.
[DNLM: 1. Accounting—United States. 2. Health Expenditures—United States. 3. National Health Programs—United States. 4. Public Health—economics—United States. W 74 AA1]
RA410.53.N396 2010
338.4'33621—dc22

2010028929

Additional copies of this report are available from The National Academies Press, 500 Fifth Street, NW, Lockbox 285, Washington, DC 20055; (800) 624-6242 or (202) 334-3313 (in the Washington metropolitan area); Internet, http://www.nap.edu.

Printed in the United States of America

Suggested citation: National Research Council. (2010). *Accounting for Health and Health Care: Approaches to Measuring the Sources and Costs of Their Improvement.* Panel to Advance a Research Program on the Design of National Health Accounts, Committee on National Statistics. Division of Behavioral and Social Sciences and Education. Washington, DC: The National Academies Press.

THE NATIONAL ACADEMIES

Advisers to the Nation on Science, Engineering, and Medicine

The **National Academy of Sciences** is a private, nonprofit, self-perpetuating society of distinguished scholars engaged in scientific and engineering research, dedicated to the furtherance of science and technology and to their use for the general welfare. Upon the authority of the charter granted to it by the Congress in 1863, the Academy has a mandate that requires it to advise the federal government on scientific and technical matters. Dr. Ralph J. Cicerone is president of the National Academy of Sciences.

The **National Academy of Engineering** was established in 1964, under the charter of the National Academy of Sciences, as a parallel organization of outstanding engineers. It is autonomous in its administration and in the selection of its members, sharing with the National Academy of Sciences the responsibility for advising the federal government. The National Academy of Engineering also sponsors engineering programs aimed at meeting national needs, encourages education and research, and recognizes the superior achievements of engineers. Dr. Charles M. Vest is president of the National Academy of Engineering.

The **Institute of Medicine** was established in 1970 by the National Academy of Sciences to secure the services of eminent members of appropriate professions in the examination of policy matters pertaining to the health of the public. The Institute acts under the responsibility given to the National Academy of Sciences by its congressional charter to be an adviser to the federal government and, upon its own initiative, to identify issues of medical care, research, and education. Dr. Harvey V. Fineberg is president of the Institute of Medicine.

The **National Research Council** was organized by the National Academy of Sciences in 1916 to associate the broad community of science and technology with the Academy's purposes of furthering knowledge and advising the federal government. Functioning in accordance with general policies determined by the Academy, the Council has become the principal operating agency of both the National Academy of Sciences and the National Academy of Engineering in providing services to the government, the public, and the scientific and engineering communities. The Council is administered jointly by both Academies and the Institute of Medicine. Dr. Ralph J. Cicerone and Dr. Charles M. Vest are chair and vice chair, respectively, of the National Research Council.

www.national-academies.org

PANEL TO ADVANCE A RESEARCH PROGRAM ON THE DESIGN OF NATIONAL HEALTH ACCOUNTS

JOSEPH P. NEWHOUSE (*Chair*), Division of Health Policy Research and Education, Harvard University
DAVID M. CUTLER, Department of Economics, Harvard University
DENNIS G. FRYBACK, Department of Population Health Sciences, University of Wisconsin, Madison
ALAN M. GARBER, Department of Veterans Affairs, Palo Alto Health Care System, and School of Medicine, Stanford University
EMMETT B. KEELER, RAND Graduate School, Santa Monica, CA
ALLISON B. ROSEN, School of Public Health, University of Michigan
JACK E. TRIPLETT, Brookings Institution, Washington, DC

CHRISTOPHER D. MACKIE, *Study Director*
MICHAEL J. SIRI, *Program Associate*

COMMITTEE ON NATIONAL STATISTICS
2009-2010

Preface

It has become trite to observe that increases in health care costs have become unsustainable. How best for policy to address these increases, however, depends in part on the degree to which they reflect changes in the quantity of medical services as opposed to increased unit prices of existing services. And an even more fundamental question is the degree to which the increased spending actually has purchased improved health.

This report addresses both of these issues. The government agencies responsible for measuring unit prices for medical services have taken steps in recent years that have greatly improved the accuracy of those measures. Nonetheless, this report has several recommendations aimed at further improving the price indexes. Because medical care is such a large part of the economy, inaccurate medical price indexes can cause significant inaccuracies in overall measures of inflation.

And accurate measures of inflation matter a great deal for policy: they affect the tightness of monetary and fiscal policy; they affect government budgets, because about a third of the federal budget is indexed for inflation; and inaccurate price indexes by definition lead to inaccurate measures of productivity.

Measuring the price of medical services well is difficult, but the ultimate question is the degree to which monies spent on medical services, as well as other policy measures, affect health outcomes. This question is much harder than measuring prices, but the panel recommends some steps it thinks will improve the nation's capacities in this domain.

This report has been reviewed in draft form by individuals chosen for their diverse perspectives and technical expertise, in accordance with procedures approved by the Report Review Committee of the National Research Council

(NRC). The purpose of this independent review is to provide candid and critical comments that will assist the institution in making its published report as sound as possible and to ensure that the report meets institutional standards for objectivity, evidence, and responsiveness to the study charge. The review comments and draft manuscript remain confidential to protect the integrity of the deliberative process. The panel wishes to thank the following individuals for their review of this report: Thomas E. Getzen, Fox School of Business, Temple University; Paul B. Ginsburg, Center for Studying Health System Change, Washington, DC; Sherry Glied, Department of Health Policy and Management, Mailman School of Public Health, Columbia University; Dale W. Jorgenson, Department of Economics, Harvard University; J. Steven Landefeld, Bureau of Economic Analysis, U.S. Department of Commerce; Mary O'Mahony, National Institute of Economic and Social Research (UK) and University of Birmingham; and Michael Stoto, Health Systems Administration and Population Health, Georgetown University School of Nursing and Health Studies.

Although the reviewers have provided many constructive comments, and improved the content of the report a great deal, they were not asked to endorse the conclusions or recommendations; nor did they see the final draft of the report prior to its release. The review of this report was overseen by Katharine G. Abraham, Joint Program in Survey Methodology, University of Maryland, and Charles E. Phelps, university professor and provost emeritus, University of Rochester. Appointed by the NRC's Report Review Committee, they were responsible for making certain that an independent examination of this report was carried out in accordance with institutional procedures and that all review comments were carefully considered. Responsibility for the final content of this report rests entirely with the authoring panel and the institution.

Many others generously gave of their time to present at meetings and to answer questions from panel members and staff, thereby helping the panel to develop a clearer understanding of key issues relevant to the development of medical care and health accounting systems. The panel especially thanks the National Institute on Aging (NIA) Division of Behavioral and Social Research, which supported the work of the panel as it wrestled over many months with the difficult issues in conceptualizing health and medical care accounts and moving toward their development, and the federal statistical agencies, which allowed the panel access to key personnel with extensive expertise about various data programs. Richard Suzman and John Haaga of NIA provided insights and guidance as project initiators. Perspectives from other interested agencies were expertly supplied by Todd Caldis, Cathy Cowan, Jonathan Cylus, Mark Freeland, Stephen Heffler, Arthur Sensenig, and Andrea Sisko of the Centers for Medicare & Medicaid Services (U.S. Department of Health and Human Services); by Ralph Bradley, John Greenlees, Michael Horrigan, John Lucier, Robert McClelland, Bonnie Murphy, and Roslyn Swick of the Bureau of Labor Statistics (U.S. Department of Labor); by Ana Aizcorbe, Dennis Fixler, and J. Steven Landefeld of the Bureau of Economic Analysis (U.S. Department of Commerce); by Jessica Banthin, Yen-Pin Chiang, Steven Cohen, and William Lawrence of the Agency for Healthcare Research and Quality (U.S. Department of Health and Human Services); by Linda Bilheimer and Christine Cox of the National Center for Health Statistics (Centers for Disease Control and Prevention, U.S. Department of Health and Human Services); by Theodore Stefos of the U.S. Department of Veterans Affairs; and by Anne Hall of the Board of Governors of the Federal Reserve.

The panel also learned a great deal from hearing about health data efforts by government agencies and researchers abroad. For their participation and willingness to travel great distances to do so, the panel thanks John Goss, Australian Institute of Health and Welfare; Sandra Hopkins, Organisation for Economic Co-operation and Development Health Division; Mary O'Mahony, National Institute of Economic and Social Research (UK) and University of Birmingham; and Michael Wolfson, Statistics Canada.

On the home front, the panel could not have conducted its work without an excellent and well managed NRC staff. In that regard, it appreciates the support of Constance Citro, director of the Committee on National Statistics; Michael Siri, program associate; and Christopher Mackie, the panel's study director.

Most importantly, I thank the members of the panel for their hard work. This report reflects the collective expertise and commitment of the individual members of the panel. All participated in the panel's many meetings and in drafting material for discussion and, ultimately, for the report itself. Each member brought a critical perspective, and our meetings provided many opportunities for panel members to learn from one another.

Joseph P. Newhouse, *Chair*
Panel to Advance a Research Program on
the Design of National Health Accounts

Contents

APPENDIXES

Acronyms and Abbreviations

ACES	Annual Capital Expenditures Survey
ACG	adjusted clinical groups
ACS	American Community Survey
ADL	activities of daily living
AHIP	America's Health Insurance Plans
AHRQ	Agency for Healthcare Research and Quality
APG	Ambulatory Patient Groups
ASM	Annual Surveys of Manufactures
ATUS	American Time Use Survey
BEA	Bureau of Economic Analysis
BLS	Bureau of Labor Statistics
BMI	body mass index
BRFSS	Behavioral Risk Factor Surveillance System
CAT	computerized adaptive test
CCS	Clinical Classification Software
CDC	Centers for Disease Control and Prevention
CDM	chronic disease model
CES	Current Employment Survey
CIHI	Canadian Institute for Health Information
CIR	Current Industrial Reports
CMS	Centers for Medicare & Medicaid Services
CNSTAT	Committee on National Statistics
COI	cost of illness

CPI	Consumer Price Index
CPS	Current Population Survey
CRG	clinical risk grouping
CRIW	Conference on Research in Income and Wealth
DALY	disability-adjusted life year
DCG	diagnostic cost groups
DoD	U.S. Department of Defense
DRG	diagnostic-related groups system
EKG	electrocardiograph
EMG	electromyography
EPA	U.S. Environmental Protection Agency
ER	emergency room
ETG	episode treatment groups
EU	European Union
FRB	Federal Reserve Board
GDP	gross domestic product
HCPCS	Healthcare Common Procedure Coding System
HCUP	Healthcare Cost and Utilization Project
HHS	U.S. Department of Health and Human Services
HIPAA	Health Insurance Portability and Accountability Act
HIV	human immunodeficiency virus
HIV/AIDS	human immunodeficiency virus/acquired immunodeficiency syndrome
HOS	Medicare Health Outcomes Survey
HPV	human papillomavirus
HRQoL	health-related quality of life
HRS	Health and Retirement Study
HUI2	Health Utilities Index Mark 2
HUI3	Health Utilities Index Mark 3
IADL	instrumental activities of daily living
ICD-9	International Statistical Classification of Diseases and Related Health Problems, ninth revision
ICD-9-CM	International Classification of Diseases, ninth revision, Clinical Modification
ICD-10	International Statistical Classification of Diseases and Related Health Problems, tenth revision
ICT	information and communications technology

IOM	Institute of Medicine
IPI	International Price Index
IRS	U.S. Internal Revenue Service
ISR	University of Michigan Institute for Social Research
IT	information technology
JCUSH	Joint Canada/United States Survey of Health
KLEMS	inputs for medical services that go into a medical care account: capital services (K); labor services—the vector of all labor inputs, from surgeons to janitors (L); energy (E); intermediate or purchased materials, which, in medical care–providing industries, includes pharmaceuticals used in hospitals and clinics (M); and purchased services (S)
LDC	less-developed countries
LE	life expectancy
LP	labor productivity growth
MC	marginal cost
MCBS	Medicare Current Beneficiary Survey
MDC	major diagnostic category
MedPAC	Medicare Payment Advisory Commission
MEG	medstat episode groups
MEPS	Medical Expenditure Panel Survey
MFP	multifactor productivity
MI	myocardial infarction
MRI	magnetic resonance imaging
NAICS	North American Industry Classification System
NAPCS	North American Product Classification System
NBER	National Bureau of Economic Research
NCHS	National Center for Health Statistics
NCS	National Comorbidity Survey
NCS-R	National Comorbidity Survey-Replication
NCVHS	National Committee on Vital and Health Statistics
NDC	National Drug Code
NDI	National Death Index
NEFS	National Epidemiologic Followup Study
NHA	National Health Account
NHANES	National Health and Nutrition Examination Survey
NHCS	National Health Care Survey
NHEAs	National Health Expenditure Accounts

NHHCS	National Home and Hospice Care Survey
NHIS	National Health Interview Survey
NHMS	National Health Measurement Study
NHSDA	National Household Survey on Drug Abuse
NIA	National Institute on Aging
NIESR	National Institute of Economic and Social Research (UK)
NIH	National Institutes of Health
NIPAs	National Income and Product Accounts
NIS	Nationwide Inpatient Sample
NMES	National Medical Expenditures Survey
NNHS	National Nursing Home Survey
NOMESCO	Nordic Medico-Statistical Committee
NRC	National Research Council
NSAS	National Survey of Ambulatory Surgery
NSF	National Science Foundation
OECD	Organisation for Economic Co-operation and Development
OTC	over the counter
PCE	personal consumption expenditures
PHC	personal health care
PPI	Producer Price Index
PPMS	Provider Performance Measurement System
PROMIS	Patient-Reported Outcomes Measurement and Information System
QALE	quality-adjusted life expectancy
QALY	quality-adjusted life year
QOL	quality of life
QWB-SA	Quality of Well-Being Scale, self administered
R&D	research and development
RRU	relative resource use
SAMHSA	Substance Abuse and Mental Health Services Administration
SEER	Surveillance, Epidemiology, and End Results Program
SEER-CMHSF	Surveillance, Epidemiology, and End Results-Continuous Medicare History Sample File
SES	socioeconomic status
SHA	system of health accounts
SID	State Inpatient Database

TFP	total factor productivity
USVEQ	U.S. Valuation of the EQ-5D
VA	U.S. Department of Veterans Affairs
VAS	visual analog scale
VBID	value-based insurance design
WHO	World Health Organization
WHS	World Health Survey

Abstract

In order for policy makers to pursue informed actions to enhance efficiency of the nation's approach to medical and health care—whether through carefully targeted cost reductions or improved performance—a redesigned data system for tracking resource productivity is needed. This report lays out strategies for advancing this objective. Specifically, the panel recommends that work proceed on two projects that are distinct but complementary in nature: the first involves reformulating the economic accounting of inputs and outputs for the medical care sector; the second involves developing a data system that coordinates population health statistics with information on the determinants of health. Though the scope of activities required for each of these two projects is different, both economic problems involve identifying units of measurement for which meaningful prices and quantities can be attached so that returns to investments in health can be estimated, tracked over time as the quality of care and the composition of the population change, and compared under alternative planning scenarios.

Inputs to medical care include capital, labor, energy and materials, research and development, and the like. The report gives considerable attention to how expenditures on these inputs are to be allocated in an accounting structure, with the panel recommending that a substantial portion can be framed in terms of treatments for diseases and other well-defined conditions. In principle, this structure allows the value of the output of medical care to consumers (patients) to be adjusted to reflect changing quality of outcomes.

Inputs to health, the output of a broader accounting concept, include medical care but also many other factors. An essential component of this kind of account—and, more immediately, a data system that could be used in its development—involves selecting a summary measure of population health, and the report assesses the options. Though it involves very long-term commitments, efficient management of health care resources requires developing a more complete understanding than currently exists of the links between population health and the array of health inputs. Thus, the report discusses data needs and issues that are confronted in research seeking to attribute health effects to both medical and nonmedical (as well as market and nonmarket) inputs to health.

Summary[1]

WHY A NATIONAL HEALTH ACCOUNT?

Health and health care in the 21st-century United States can be partially characterized by a few indisputable trends. The country is devoting a large and growing share of its resources to medical care, spending $2.3 trillion, or 16 percent of gross domestic product (GDP), on it in 2007. That figure is projected to rise to perhaps $4.2 trillion, or 20 percent of GDP, by 2016 (Poisal et al., 2007). Not only is this spending massive in an absolute sense, it is also high comparatively, as the median expenditure of Organisation for Economic Co-operation and Development countries is around 8.5 percent of GDP. The budgetary pressures being created by this allocation—at individual, organizational, and national levels—have led to growing criticism of the productivity of U.S. health care. As the population continues to age, the strain that will be put on publicly funded programs such as Medicare will only elevate the debate about how best to meet the nation's health care needs.

One factor contributing to the trend of rising expenditures is the ever-expanding and increasingly sophisticated array of life-extending and life-enhancing treatments produced by medical research for the range of conditions and diseases that afflict the population.[2] In addition, the body of knowledge about the effect of

[1]This summary includes mainly top-level recommendations; the body of the report contains additional detailed recommendations, particularly with respect to data needs.

[2]A recent study of the cholesterol-fighting statin Crestor—which enrolled 17,802 subjects in 26 countries—illustrates the point. The study found that the drug could cut cardiovascular risk in those with normal cholesterol by 45 percent (see, e.g., Ridker et al., 2008). According to analysis by James Stein of the University of Wisconsin School of Medicine and Public Health, U.S. physicians might

nonmedical factors—such things as diet, exercise, and environment—on people's health continues to grow.

For all these advances, a gaping hole in information still exists. Relative to knowledge about health care expenditure and medical science, much less is known about the return that individuals, and society in general, receive for the investment in health. Given the massive amount of resources that are spent publicly and privately, it is astonishing that so little effort has been made to systematically assess what we are buying for this investment.

At the heart of this information chasm is the need for data on how inputs into medical care translate into outputs—completed treatments and procedures—that, in combination with other factors, ultimately affect the population's health. While some studies have suggested that productivity growth in medical care is reasonable in aggregate (Cutler and McClellan, 2001; Cutler, Rosen, and Vijan, 2006), others argue that there is substantial waste at the margin (Fisher et al., 2003; Skinner, Staiger, and Fisher, 2006). In this report, we provide guidance about what data are needed to measure the outputs produced by the medical care sector. Without this kind of information, it is impossible to credibly assess whether the nation spends too much or too little on medical care relative to, say, public health measures, and, perhaps more importantly, whether we purchase something close to the right mix of medical care goods and services for a given level of resources expended. In order for policy makers to pursue actions that reduce costs sensibly, improve performance, and, in general, enhance the efficiency of the national approach to health and medical care, a more systematic approach to compiling data for the purpose of tracking productivity in the sector is needed.

The National Income and Product Accounts (NIPAs) produced by the Bureau of Economic Analysis (BEA) and the National Health Expenditure Accounts (NHEAs) produced by the Centers for Medicare & Medicaid Services (CMS) are the foundational components of the U.S. health care data infrastructure. While the virtues and utility of these data sources are well known, they are not sufficient on their own to inform policy. The national accounts have particular difficulty decomposing medical spending increases into price, quantity, and quality change elements. For example, an increase (or decrease) in the observed price of treating a disease may reflect a change in the price of unchanged treatment inputs, a change in the amount of inputs (e.g., a surgeon's time) required, the development and use of new drugs or procedures that alter outcomes, or simultaneous changes in more than one of these factors (National Research Council, 2005, pp. 117-118).

In this report, we offer guidance for extending these important data sources, at first in an experimental or satellite account setting, to better inform national decisions about resource allocations to health care. A health data system built on

---

respond by prescribing statins to as many as 7 million more patients—at a cost of about $9.7 billion annually (Light, 2008).

economic accounting principles has the potential to provide a comprehensive picture of population health in relation to health care spending within an integrated framework in which consistent definitions, measurement tools, and analytic conventions are used (Rosen and Cutler, 2007). Economic accounts provide the key data infrastructure for measuring productivity, which, in turn, can be used to identify high-return spending and investment areas.

Given the wide range of statistical and research measurement needs, creating health data systems that will adequately inform policy requires a multipronged effort. Economic accounting (as practiced by BEA), price and productivity measurement (Bureau of Labor Statistics [BLS]), medical and health-oriented research (e.g., funded by the National Institute on Aging [NIA], National Institutes of Health), and health monitoring (e.g., National Center for Health Statistics, [NCHS], Centers for Disease Control and Prevention [CDC]) are all data driven, but requirements vary in terms of needed characteristics and complexity. BEA's top priority is to measure medical care inputs and outputs correctly, since this is a major component of the market economy; the needs of BLS are similar. However, many health policy researchers are interested in establishing links between population health and its determinants—medical and otherwise—in order to develop strategies for improving care in a cost-effective manner. These efforts demand data on the multitude of factors affecting population health and on health status itself.

Because of the nation's need for more accurate information on its health care system costs and benefits, it is essential to improve the data infrastructure.

Recommendation 1.1: Work should proceed on two projects that are distinct but complementary in nature. One accounts for inputs and outputs in the medical care sector; the other involves developing a data system designed to track current population health and coordinate information on the determinants of health (including but not limited to medical care).

Groundbreaking work that will create a foundation for the first type of account is indeed already under way at the statistical agencies. Research programs that will contribute toward fulfilling the second objective are still in their infancy (and taking place predominantly in academic settings) and will take much longer to mature. Ideally, though, the national health data system should eventually be maintained by the statistical system.

WHAT KIND OF HEALTH ACCOUNT?

Constructing a fully developed set of national health accounts involves three broad steps, which define the topics of interest in this report: (1) compiling detailed and comprehensive data covering the nation's expenditures on and utilization of health care, organized in such a way that prices and quantities of meaningfully defined units of goods and services can be measured; (2) assessing summary measures that allow the changing state of the population's health to be tracked along a number of dimensions, and identifying those that are most appropriate for use in a health accounting framework; and (3) developing an integrated data system that allows researchers to investigate links between medical interventions and other health-related activities on one hand and population health on the other.

At the outset, we recognize (and consider it a positive) that work to improve health data within a national accounting framework is moving forward along several fronts and is motivated by multiple objectives. One strain of research is focusing on the inputs and outputs of the medical sector. Here, as would be consistent with the market-oriented NIPAs, units of medical care produced (however defined) can rightly be considered the output. This work is of immediate concern to statistical agencies such as BEA, BLS, and CMS that are responsible for producing expenditure, price, and productivity statistics.

A second strain of research is oriented toward developing statistical data for relating the population's health status to an array of factors that affect that status in a kind of health production function. Inputs include medical care, time spent by people investing in their own health (e.g., exercise), consumption items (e.g., food), research and development, and the quality of the environment. The output—better health—produced from investments in these inputs includes both length and quality of life dimensions, which can be conceptualized jointly in summary measures of population health. In the broad-concept account, medical care is an input in the production of "health"—treatments are pursued with the objective of extending or enhancing a patient's years of life.

The two projects are complementary; the analytic pieces for which the statistical agencies are responsible—e.g., medical care expenditure accounts and price indexes—are intermediate building blocks for a comprehensive health data system. Such a system would in turn provide a foundation for developing a national health account, such as that envisioned in the report *Beyond the Market: Designing Nonmarket Accounts for the United States* (National Research Council, 2005). An initial task for both lines of work is to accurately categorize nominal expenditure estimates into meaningful production units to which prices and quantities can be attached. Even those topical areas for which the statistical agency and research community objectives appear to diverge may do so for only a period. For example, for some time, researchers have investigated the health impact on the population of various disease treatments produced by the medical sector; at the moment, this is not an active area for BEA. However, ultimately, both BEA (NIPAs) and BLS (price index work) should be interested in integrating medical outcome data into their analytic apparatus in order to track changes in the quality of service, which, along with the quantity, determines the value of the industry's output. The most methodologically advanced components of the NIPAs, and of the various price indexes and productivity measures, already track the changing quality of goods and services and facilitate the ability to break out nominal price changes into real price and quality change components.

In thinking about next steps of these research programs, we argue that, among the inputs to health, the medical care piece should initially be given priority. For policy purposes, the most pressing need is to measure medical care expenditures and related outputs and outcomes as well as possible. In addition, accurate expenditure and price data on medical care are essential both for developing broader health accounts and for improving the medical care component of the NIPAs. Thus, the panel supports the idea that, at least initially, BEA's satellite health accounting program should focus on input and output data relevant to the provision of the nation's medical care.

The satellite account should include all inputs to the production of medical care, whether they are purchased by the medical "industry" or by households and regardless of how directly they affect health. In addition to clear cases of life-saving medical interventions, treatments that improve quality of life (e.g., knee repair), that are preventive in nature (e.g., a routine physical), or even that have ambiguous links to health (e.g., some over-the-counter medications) should be included so long as they are part of the medical care sector. For setting account scope, we deemphasize concerns about the extent to which the procedures improve health (physical or mental). Ideally, however, outcomes research will indicate quantitatively—in an accounting framework that values system output in terms of improvements in quality or quantity of life—which medical services are more and less productive. Arguments in favor of this approach are that it creates a relatively clear line of demarcation and that it maintains historical comparability. Also, for policy, it is important to know expenditure totals for the sector and to have an idea what people are getting for that spending.

A TREATMENT-OF-DISEASE-BASED FRAMEWORK

In building a broadly useful national health accounting system, the question of how to define the unit of measurement must be tackled. Ideally, medical sector goods and services would be defined in such a way that (1) expenditures could be estimated each period for every good or service produced by the industry, (2) meaningful quantities and prices (nominal and real) could be tracked, and (3) quality change of the goods and services could be monitored. One way to proceed that embodies these three goals is to identify the output of the medical care sector as completed treatments or procedures.

A treatments-based organizing framework coordinates logically with a broader health data system because, in principle, it creates a unit of analysis for which changes in the effectiveness of various medical services may be monitored. It provides a mechanism whereby prices can be adjusted to reflect changing quality, the substitution of inputs can be handled better than they are currently, and the introduction of new treatments can be dealt with on a disease-by-disease basis. At the present time, however, health expenditure and outcomes data are not organized in a way that feeds naturally into this accounting framework, and so their structure needs to be modified.

Although the logic of a treatment-of-disease approach is clear, it will not be easy to execute in a comprehensive national data system. It requires defining episode groups and then dealing with population heterogeneity within condition categories. A first step is to agree on a crosswalk between existing classification systems and a version that is at the right level of detail for a national health account—one logical option would be an aggregation of International Classification of Diseases-Tenth Revision categories although the final version might be arranged quite differently. Even then, problematic conceptual issues associated with using a treatment-of-disease approach remain, such as those created by the presence of comorbidities. A reasonable starting point is to treat common clusters of comorbidities (e.g., diabetes and vascular complications) as distinct disease categories; in addition, methods exist for allocating expenditures on patients with multiple conditions based on ratios of costs estimated from patients with single diagnoses. A special advisory committee could be convened by BEA to provide guidance on these conceptual and definitional tasks.

Another hurdle to implementing a disease treatment approach is the problem of defining a measurement unit for chronic episodes or any kind of service that involves a long treatment period. In the literature, episode-based price indexes have defined the good as a completed episode (e.g., the price of treating a heart attack). At the end of that episode, expenditures on the treatment are collected and that is the price of the unit. For the purpose of pricing the treatment of chronic illnesses, which may span several statistical reporting periods, the unit of measurement should include all spending on the disease for some designated and consistent period of time, most likely a year.

NONDISEASE-SPECIFIC MEDICAL CARE

A large portion of medical care activities, in terms of costs, can be accounted for by looking at treatments of specific diseases. Roehrig et al. (2009) produced estimates of national health spending by medical condition using 260 categories defined in the Agency for Healthcare Research and Quality (AHRQ) Clinical Classification Software, which groups the numerous International Classification of Diseases-Ninth Revision codes into broader categories that are "clinically meaningful." Reallocating the NHEA totals using data from the Medical Expenditure Panel Survey (MEPS), the research team found that the largest eight expenditure categories (circulatory system, mental disorders, musculoskeletal, injury and poisoning, digestive, neoplasms, respiratory, and nervous system) accounted for almost 70 percent of total expenditures. By contrast, prevention, exams, and dental—which would not fit cleanly into disease-specific categories—accounted for around 6 percent of personal health spending; another 6 percent of total personal health expenditures from the NHEA went unallocated.

Although disease treatment categories are clearly important, it is obvious that not all medical services (and certainly not all health-related services) can be captured in them. Due to the presence of these other aspects of care, including those that may take place before and after treatments, it would not be desirable to shoehorn all medical care spending (and associated health effects) into disease categories. These areas of care might be better considered separately. We envision that, ultimately, some catch-all categories will need to be created for nondisease-specific health care and also for episodes of management—such as screening, diagnosis, and prevention.

Recommendation 2.11: Although starting with medical care on a disease-by-disease basis is a realistic way to proceed in order to begin accounting for a very significant share of the medical care economy, work should also begin on estimating the costs of, and eventually the health return from, interventions other than treating specific diseases (e.g., management, preventive, diagnostic, screening) and long-term medical services.

The idea would be to track quantities and prices of items in these categories over time, just as one would track direct disease treatment. In some cases, it may be possible to link diagnostic tests and the other forms of care backward from an eventual resulting treatment. The task of setting up the nondisease treatment categories could again be assisted by an advisory committee convened under the auspices of some combination of CMS, AHRQ, and NCHS, although one office will need to take the lead.

ALLOCATING SPENDING ACROSS TREATMENT-OF-DISEASE CATEGORIES

The first major task in developing a national health account—whether it is BEA's satellite version or the broader health type—is to devise a method for allocating economy-wide spending on medical care into the treatment-of-disease categories described above. Because they serve as building blocks for many kinds of health data systems, improving the methodology for organizing and tracking health care expenditures is an immediate priority.[3]

Given the number of different disease classification schemas currently in use in the U.S. health care system, it is essential for all players, in both public and private sectors, to participate in and come to a consensus on the development of a single unified version.

[3]Of course, a parallel account can also be maintained to serve needs that rely on the current NHEA structure.

Recommendation 3.1: A concerted effort is needed to reach consensus on how to classify diseases and about what the criteria are by which diseases are disaggregated from the very broad International Classification of Diseases chapters. The National Center for Health Statistics should lead the effort, working with the Agency for Healthcare Research and Quality, the Centers for Medicare & Medicaid Services, the Bureau of Economic Analysis, and other relevant statistical agencies. As part of this effort, U.S. agencies should participate in ongoing standardization efforts (such as those sponsored by the Organisation for Economic Co-operation and Development or the World Health Organization) to benefit from international expertise, to consider these as the basis for a national system, and to facilitate international comparisons.

Once an agreed-on classification system has been established, several options exist for attributing spending across treatment episodes for the range of disease categories. One is an encounter-based method in which spending is attributed to one or to several diagnoses as reflected by data extracted from patient claims. A second, broader approach involves constructing episodes of treatment—which may include numerous encounters over a predefined period—then adding up dollars spent nationally on each of the range of diseases and conditions. A third possibility is a person-based approach, in which spending on various treatments is tracked on a person-by-person basis over a set period of time.

BEA, for its part, is working both internally and with outside researchers to establish what the allocations look like under the different methods and whether it matters for estimating aggregate expenditures and prices. These researchers (Rosen and Cutler, 2007) have already begun producing episodes-of-treatment cost estimates for the different methods, demonstrating that it can be done; spending could also be further broken down into additional subcategories such as disease prevention, diagnosis, and screening activities. Whichever method of allocating expenditures is chosen, it has to offer a solution to the comorbidity problem.

At this point, the panel cannot definitively recommend one method for tracking expenditures and output (and eventually outcomes) associated with medical treatments over all others. And the appropriate concept depends a great deal on the specific application. For example, if the goal is to compare costs and health effects for a given disease, as is done in cost-effectiveness analyses, a person-based approach is likely to be most appropriate. In contrast, if price index construction is the goal, federal agencies may find an episode-of-treatment approach more meaningful. For managers of a health plan trying to understand why emergency room spending patterns have changed, real-time answers may be possible only with an encounter-based approach. What is needed is more empirical work to compare different approaches and to determine more definitively which is best under different conditions.

Recommendation 3.3: The Bureau of Economic Analysis (working with academic researchers and with the Bureau of Labor Statistics) should continue to investigate the impact of different expenditure allocation approaches—particularly the episode- and person-based methods—on price index construction and performance. Research is needed to determine which method is best under different circumstances.

As part of this effort, BEA (perhaps in coordination with AHRQ and NIA) should sponsor a workshop during which private-sector vendors that produce various disease-grouping systems could present their products, identify how they are used in the marketplace, and describe the underlying rules and logic.

At this point, a cautious approach is warranted, as it is too early for BEA to buy into a particular method for aggregating treatment costs. This means that there may need to be parallel sets of accounts going on, at least on a research basis, for some time. The statistical agencies should continue to experiment with competing methods, and research should be designed to test variation in results from different data sources.

Recommendation 3.4: The Bureau of Economic Analysis, working with academic researchers (and perhaps other agencies, such as the Centers for Medicare & Medicaid Services and other parts of the Department of Health and Human Services), should collaborate on work to move incrementally toward the goal of creating disease-based expenditure accounts by attempting a "proof of concept" prototype. Using a subgroup of the population with good data coverage, the prototype would attempt to demonstrate that dollars spent in the economy on medical care can be allocated into disease categories in a fashion that yields meaningful information.

The Medicare population, the military, or veterans—groups for which there is available spending and health data (and for whom a lot of the medical care action takes place)—would be logical choices. Alternatively, a disease-costing pilot could be done for a well-defined, geographically (and administratively) complete group, such as found in parts of Intermountain Healthcare, the Geisinger Health Care System, or one of the Hawaiian islands before attempting it on a national basis.

DEVELOPING PRICE INDEXES FOR MEDICAL CARE INPUTS AND OUTPUTS

Much of what is required to develop health and health care accounts has to do with medical care price deflation. Once the nominal expenditure flows for the sector have been figured out correctly, a daunting problem in itself, the next task is to begin estimating disease-based price indexes. In this report, we take the position, expressed by many other health economists, that price indexes organized by treatments of disease or episodes of illness should be used as deflators for the medical sector components of the national accounts.

Health economists have developed the conceptual tools that are needed to construct price indexes for medical care goods and services demanded by consumers. Much of the academic literature has relied on patient claims data to provide a picture of price trends for treating specific conditions (e.g., Berndt, Busch, and Frank, 1998). In these studies, the goods and services have been defined as a completed episode; this parallels the conceptual approach described above on how to categorize nominal expenditures. For example, for a heart attack patient, this may involve time and expenditures on a series of initial treatments plus those that take place during the recovery period. At the end of that episode, data are collected to estimate all dollars spent over the entire period; this forms the basis for pricing a completed episode.

Under optimal conditions, a price index will embed a capability to pick up the substitution of inputs that takes place over time. For its satellite medical care account, BEA proposes to take the system-wide spending over some period of time on treatments of each condition (such as for depression), regardless of treatment mode (e.g., drug versus talk therapy), and divide it by the number of patients treated. The idea is to calculate a unit value that includes all spending and then allows for substitution across treatment types as well as for the introduction of new drugs or other therapies.

A number of examples can be cited from the literature of how spending by traditionally defined medical care industries is handled to form the price of a specific treatment. Pricing of cataract treatment is one of the most prominent of these (Shapiro, Shapiro, and Wilcox, 2001). If price indexes track industries in a way that segregates by type of service facility, then surgeries taking place in a hospital are sampled to generate one price index, whereas surgeries taking place in clinics are sampled in the estimation of another. A traditional industry-based index of this sort will miss the price effect that accompanies a scenario in which people switch from the more expensive hospital surgery to a less expensive ambulatory care facility or an office. Assuming quality remains constant (i.e., treatment outcomes are comparable), then a direct comparison price index capable of capturing the facility shift would more accurately estimate the decrease in price.

Even though the BLS Producer Price Index (PPI) for hospitals has been available with a cost-of-disease classification for more than 15 years, BEA does not deflate (adjust for price inflation) at the disease treatment level. Expenditure data are not available by cost of disease, so there is nothing for BEA to deflate. For that reason, BEA uses the aggregate PPI for hospitals as the deflator. Clearly, the units of analysis should be the same for price and expenditure data used in a medical care account (or any economic account for that matter). In the future, this problem could be resolved by using new data from the Census Bureau and from the PPI, and the U.S. statistical system will have gone a long way toward providing expenditures and price indexes for medical care that are harmonized around a cost-of-disease framework.

Recommendation 4.3: The Census Bureau should give high priority to providing annual data for hospitals and other medical care industries, grouped by a cost-of-disease system that matches the one used in the 2007 Economic Census and in the Producer Price Index. The panel notes the welcome provision of new data on receipts categorized by disease that accompany publication of the 2007 Economic Census, though a considerable amount of work remains to be done to bring the quality of these data up to the standards needed for use in official statistics.

QUALITY CHANGE

As a first step to improving price measurement for medical care, it is useful to account for the reduction in costs resulting from the substitution of inputs in treatments. However, in order to tell the full story, we must turn to the (probably more quantitatively significant) question, "Is the new treatment technology better, worse, or the same as the old one?" Although BEA's plan is to defer some aspects of price work until a greater consensus about methods emerges, it is essential to think about quality change from the beginning. Improvement in medical procedures creates a major measurement issue, and any price index that does not confront it will ultimately be less than satisfactory. BEA is aware of the importance of quality adjustment but, at the moment, reasonably claims to not have a systematic way of dealing with it for medical care. Perhaps, however, there are some cases that can be addressed on the basis of the existing literature and that BEA could get started on sooner rather than later.

Thinking seriously about how to measure changes in the quality and, in turn, the real cost of medical care requires the monitoring of information about outcomes associated with that care. These efforts will require expertise from medical researchers. Information about medical care outcomes and the changing quality of treatments is also essential for balanced policy discussions about resource allocation, which have focused nearly exclusively on the cost side while neglecting the changes in value that accrue as a result of the expenditures. As an example, the price of treating heart attack patients has increased dramatically, but so too has the desirability of outcomes. Fifty years ago, heart attack patients were told to rest—a practically free treatment for cases in which that rest was done at home. Treatments now exist that, while much more expensive, most would consider preferable because they have greatly extended life expectancy (rest is now thought to be counterproductive). When measuring prices of treating heart attack patients, the changes in the efficacy of the treatment must be monitored along with its cost. This kind of quality adjustment is a well-established component of accurate price measurement.

are complementary because of the trade-off that exists between sample size and representativeness. Survey data are essential to the accounting project because of the detailed patient information they can provide. However, their sample sizes are adequate only for high-prevalence conditions such as cardiovascular disease and risk factors. In contrast, insurance claims data provide a large sample, but at the expense of representativeness—no single source provides a national sample.

1

Health Data and Health Policy

1.1. THE PURPOSE AND VALUE OF AN EXPANDED SYSTEM OF NATIONAL HEALTH ACCOUNTS

According to the Centers for Medicare & Medicaid Services (CMS), in 2007 the United States spent $2.2 trillion—or $7,421 per person—on medical care, which accounted for 16.2 percent of the gross domestic product (GDP) for the year. As a proportion of GDP, medical care spending in the United States substantially exceeds that of other industrialized countries: among all Organisation for Economic Co-operation and Development (OECD) countries, the median proportion is 8.5 percent. Are the benefits from this magnitude of medical care spending worth it? This question is one of today's major economic issues.

Medical care spending in the United States is not only large, but it is also growing rapidly—up by 56 percent since 2000, it has roughly doubled since 1996. Increasing health care costs pose political questions that are heightened because the U.S. government pays a substantial portion of those costs in the form of Medicare, Medicaid, and other publicly funded programs. Indeed, growing health care costs are a public policy issue in all industrialized countries and, in most of them, governments finance more of the nation's health care than is the case in the United States.

What are the sources of increased medical care spending? What are the consequences of reigning in its rate of growth (if it can be done)? These vital questions must be answered in order to forge a consensus on public policy toward health care costs; however, they demand data to enlighten debate on them. The United States has aggregate data that chart its total medical care expenditures and rate of growth. The most widely cited estimate of total health care costs comes from the National Health Expenditure Accounts (NHEAs), compiled by CMS. An alternative estimate of the total can be extracted from the National Income and Product Accounts (NIPAs) that yield information on GDP and its growth and composition. The national aggregate tells us little that is useful for addressing the major questions listed above. Knowing that total medical care spending is growing is only part of the story. We also need to know if this growth is producing benefits that are keeping pace and are, in some sense, worth it. Knowing that expenditures are growing is not the same thing as knowing why they are growing.

At a highly disaggregated level, a great amount of survey and administrative data also exist that are useful for analyses of medical care, health status, and medical expenditure trends. Missing, however, is a comprehensive analytical database lying between the two extremes—one that permits analysis at a national level of the forces driving medical care costs, of growth in medical services, and of the benefits received from expenditures on health. The panel believes that a combination of improved economic accounts for medical care and a new database for analyzing the health care sector and its impacts on population health is essential for informing policies in these areas; microdata—survey and administrative records—will also always be necessary for understanding of medical care spending, health status, and costs.

Recommendation 1.1: Work should proceed on two projects that are distinct but complementary in nature. One accounts for inputs and outputs in the medical care sector; the other involves developing a data system designed to track current population health and coordinate information on the determinants of health (including but not limited to medical care).

Groundbreaking work that will create a foundation for the first project—the medical care account—is indeed already under way at the statistical agencies. For the second project, measures of current population health could be developed quickly, and information on the determinants of health could be collected more aggressively by the statistical system immediately. However, using the health data system to construct a national health account, as outlined in *Beyond the Market* (National Research Council, 2005), that systematically tracks population health and its determinants requires much more research on the impact of health care on outcomes, which determines its value. The two kinds of data systems are summarized in the sections that follow and elaborated in greater detail in Chapter 2.

1.1.1. An Economic Account for the Medical Care Sector

An economic account integrates data from diverse sources within an overall analytic framework, so that each individual entry is consistent with all the others and with the total. The best known examples of economic accounts are the NIPAs, compiled by the Bureau of Economic Analysis (BEA), and those produced by the statistical agencies of countries around the world. In the NIPAs, consumption, investment, government spending, and net foreign trade aggregate to GDP; at the same time, labor compensation, corporate profits, and rental and other incomes aggregate to a total flow of income, which balances with the flows on the expenditures side. Any component of the accounts can be used for analysis in conjunction with any other component and with any total or subtotal in the system (linking consumer spending on home electronics to household disposable income, for example). An economic account provides an integrated system in which the data can be arrayed according to the questions asked.

The existing NHEAs provide integrated information on health care spending. Although they serve a number of program and research needs, they are not designed to permit analysis of critical questions concerning medical care costs. The NHEAs provide data on the sources of financing of the medical care system (government, insurance companies, etc.) and on the institutional units—or industries—that receive funds (such as hospitals, doctors' offices, pharmaceutical providers). The NHEAs do not, however, provide any information on what is purchased with those expenditures. They are not disaggregated to show the amounts spent for treatments of diseases or other health impairments, nor are the trends in spending disaggregated along treatment, or cost-of-disease, lines. By analogy, it is as if the NIPAs presented information on consumer expenditures at department store and grocery store industries, without any information about spending patterns for clothing and furniture or for meat and fresh vegetables. Consumers do not go to grocery stores to buy grocery store visits—they buy meat and potatoes; similarly, medical insurers and others who fund medical care do not buy hospital visits, they buy treatments for heart attacks and broken legs.

Price indexes for hospitals, constructed by the Producer Price Index program of the Bureau of Labor Statistics (BLS), are already collected and published using a cost-of-disease classification, the only workable framework for obtaining price information about the cost of medical care (see Chapter 4). What has so far been missing from the NHEAs (and from the NIPAs) is a disaggregation of health care expenditures along the same lines. It is a major shortcoming in national statistics that health care, which absorbs 16 percent of GDP, is the only sector of the economy in which price indexes exist for deflation (at least for hospitals) but no comparable expenditure data exist to deflate.

Moreover, even though the NHEAs contain an investment component and data on spending on pharmaceuticals, they do not present a systematic accounting for all the inputs into the production of medical care—skilled and unskilled labor, services of capital equipment, such as scanners and so forth, pharmaceuticals and other inputs, including energy—arranged in such a way that they can be linked to the output of the sector. That is because the NHEAs are conceived as only an accounting for the sources of medical funds and the recipients of funds, not as an account that links the inputs to medical care with the output of the sector.

Thus, the first need for analysis of medical care spending is to construct an account for the medical care sector that matches what is available for other comparably sized sectors of the economy. It would link medical inputs to the output of medical care and would be parallel to the account for any other industry in BEA's industry accounts (e.g., electronics and computer production, legal services). BEA's industry accounts are an adjunct to the main accounts for GDP; they disaggregate the total economy into expenditures by industry, and they also link the outputs of industries with their respective uses of inputs. The medical care account's major requirements are comprehensive measures of the inputs to the production of medical care (including pharmaceuticals, skilled professionals, and high-tech capital equipment) and a measure of the output of the sector. It would resemble an enlarged and improved version of the account for the health and social services industry that is already in the current BEA industry accounts.

Ultimately, as described in Chapter 2, a number of improvements in BEA's industry accounts will be necessary. Perhaps the major change from the current statistical data on medical care is the disaggregation of total medical expenditures along cost-of-treating-disease lines. The formidable problems in doing so are subjects of Chapter 3. However, we emphasize that a long history of cost-of-disease accounts, dating at least to Rice (1966) and extending to Hodgson and Cohen (1999) and others, indicates overwhelmingly that construction of such accounts is feasible. Such accounts have also been estimated in a growing number of OECD countries, including Australia, Germany, the Netherlands, and the United Kingdom (Organisation for Economic Co-operation and Development, 2009). In Canada, they are now a part of the country's medical care information system. Methods for allocating expenditures to the International Classification of Diseases categories have improved over time to the point that, in recent studies, over 90 percent of spending has typically been accounted for (see http://www.rti.org/files/COI_Reviews.pdf for a review of recent cost-of-illness studies and estimates). This is despite allocation difficulties involving diagnostic expenditures, for dealing with activities associated but not clearly located in a particular disease (like ambulance services or long-term care), items that affect many diseases (like the sale of aspirin), and various overhead costs for facilities and equipment. That cost-of-disease accounts have been estimated so frequently does not mean that improvements are not needed (see Chapter 3). But they are not untried exercises, as they are sometimes mistakenly perceived to be. The challenge is to improve on what has been done before and to integrate that work into an account that provides information on cost trends over time.

Disaggregating health care expenditures along a cost-of-disease framework would be useful for addressing a range of questions: Is spending increasing at a more or less constant rate across all diseases? If not, which disease categories have the fastest increases and which the slowest? Does evidence corroborate that disease categories whose treatments involve the most rapid changes in technologies have the fastest rates of growth in expenditures, as is often alleged (but infrequently backed up with solid data)? Do changes in expenditures by disease categories correlate to changes in incidences of diseases, or are trends in the population's underlying medical status not the primary drivers of health care spending? It is not possible to design effective policies toward health care costs without being able to address such questions, yet at present no aggregate national estimates can be brought to bear on them, and much of what is known consists of partial and incomplete information and a great deal of more or less informed anecdotes.

Current information is also inadequate for ascertaining whether expenditure growth in treating diseases comes from an increase in treatments (that is, an increase in medical services), from an increase in the prices of those services, or from both. The science and technology of medical research is producing an ever-expanding and increasingly sophisticated array of life-extending and life-enhancing treatments for the range of conditions and diseases that afflict the population. Do the prices of treatments rise with increasing use of high-tech medical procedures and equipment, as sometimes alleged, or are expenditures rising because those advances produce increases in the quantity of medical services demanded and used? The expenditures in the medical care account need to be partitioned into price (medical inflation) and quantity (growth in medical services) terms, as is done routinely for accounts covering other parts of the economy.

For a medical care account, we argue that output can be defined as a completed treatment for a specific condition, or an episode of care.[1] The quantity of medical services so defined is the total output of the medical sector, and in line with the previous discussion, this output is multidimensional. Medical services arrayed by disease treatments provide the "product detail" classification for the sector that is parallel with product detail routinely provided by the statistical system in other sectors, such as electronics and computer production.

In other sectors of the economy, the usual method for estimating growth in quantities of goods and services is to "deflate" expenditure change, at a detailed level, by a price index. However, price indexes (measures of inflation) in medical care have long been considered problematic, and although substantial improvements have been made in recent years, much remains to be done. We review the problems of estimating price indexes for medical care in Chapter 4. Alternative quantity-based methods are attractive in countries where the government provides medical care directly (so the price charged is not relevant), but they are also worth considering for the United States because the usual practice of deflation by price indexes has both conceptual and practical disadvantages in the case of medical care.

The "health" account summarized above takes as its boundary the medical care sector of the economy. In terms of a distinction usually drawn in the literature, it is an account for activities that lie within the market sector, though perhaps not

[1]Hornbrook, Hurtado, and Johnson (1985) define the term "episode of care" as the "inclusive set of services provided by any and all health care providers—physicians, dentists, physician assistants, nurse midwives, nurses, chiropractors and the like—in the treatment of a given health care problem. Subcategories of an episode include disease episodes, illness episodes and health maintenance episodes."

entirely so. For example, time used for the care of the ill by volunteers, friends, and relatives could well be considered within the scope and included in the account's inputs without changing its basic character. In the medical care sector account, the output is the treatment of diseases, plus ancillary activities that contribute to control of disease, like periodic medical checkups and diagnostic services.

1.1.2. A Data System for Economic Analysis of the Production of Health

The medical care sector does not produce health; it produces medical services. The distinction between the production of medical services and the production of health is an essential one that not only has several implications, but that also suggests an inherent limitation in the type of account discussed above. Even if one were able with an improved medical care account to partition the growth in medical care expenditures into price change (inflation) and quantity change (growth in medical services) components, it still may not be possible to judge whether or not the quantity changes are in some sense excessive: Is too much being spent on late life-stage surgeries relative to preventive measures? Is too much being spent on knee replacements relative to weight reduction? Or, even, is too much spent on medical care and not enough on public health measures? Just knowing the rate of advance of medical services, arrayed by disease treatments, is not the end of the story. The big question remains: Is the $2.2 trillion the United States is spending on health care worth it? And that translates into a series of questions of the following type: Is the amount the United States spends on circulatory diseases, for example, too much, too little, or about right in terms of what we get for the dollars expended?[2]

Addressing what society and even what individuals get for their medical expenditures is complicated because the output of the medical care sector is not health. Instead, health is "produced" (in the language of economics) in conjunction with many other variables, or inputs, in addition to medical care: diet, environmental factors, lifestyle, and so forth. Changes in a nation's health may owe as much or more to those nonmedical determinants as to medical care services, a relationship that has been well articulated in epidemiology since at least McKeown (1976).

It may seem plausible that increasing the resources put into health care should yield improvements in health at a nationwide level (so the United States,

[2]The kind of detailed cost-benefit analysis required to answer such a question has permeated policy actions in some countries more deeply than it has in the United States. For example, in the United Kingdom, the National Institute for Health and Clinical Excellence determines what surgical procedures, screening tests, and drugs the national system will pay for—a drug that is predicted to add an extra six months of good-quality life for a patient for $15,150 or less is automatically approved; one adding six months for $22,750 or less might get approved; and more expensive medicines have only rarely been approved (see http://www.nytimes.com/2008/12/03/health/03nice.html?pagewanted=1&th&emc=th).

which has the world's largest medical care expenditures per capita, should have the world's best health); however, a tight relation between expenditures on medical care and the level of health does not necessarily exist, in part because of the influence of nonmedical determinants of health.

The nation therefore also needs a broader data system than that provided by the economic accounts for medical care, one that could be used to relate changes in the nation's health to the factors determining health. As with economic accounts, this data system would be designed to facilitate investigations of the link between output (health status) and inputs (which are all the influences on health, including medical care). On the output side, growth consists of changes in the population's health itself, and completed medical treatments are inputs to this production process. On the input side are statistical data on the array of factors—including, but not limited to, medical care—that affect the population's health status. This second kind of economic data system is a long-term project that is much more on the research frontier.

The medical care account and the broader health data system are linked, and work on the former is an integral part of, and logical first stage project in the production of, the latter. Moreover, these data and analytic systems could logically be constructed along disease lines, which underscores the vital role that cost-of-disease accounting (Chapter 3) plays in each. Accordingly, construction of the health data system requires data from the medical care account and work on the two needs to proceed in a coordinated fashion.

Constructing a comprehensive health data system is an ambitious undertaking. The first major empirical difficulty is measuring the output—health—a task that requires going well beyond standard longevity figures and death rates to consider morbidity. In Chapter 5, we review summary measures of health and the state of research on health measures, and consider the research issues that must be resolved if they are to be useful in an economic accounting framework.

In Chapter 6, we discuss approaches for attributing the output (improved health) level achieved to the inputs or determinants of health—medical care as well as nonmedical and nonmarket inputs to health. Measurement of the inputs that determine health also poses serious challenges. Much still remains to be learned about the nonmedical determinants of health. Fortunately, the body of knowledge about the effect of these factors—things such as diet, exercise and environment—on the population's health is growing.

1.2. REPORT AUDIENCE, REPORT STRUCTURE

The Panel to Advance a Research Program on the Design of National Health Accounts was assembled to examine how health and medical care data systems might be designed and implemented within a national accounting framework. The panel was charged to study and make recommendations about where to target research to best improve the knowledge base for developing statistical data for relating the population's health status to an array of factors—including, but not limited to, medical care—that affect that status. During the study, the panel confronted a number of vexing conceptual issues that are integral to both (1) improving input and output accounting of the medical care sector, and (2) developing broader health accounts and data systems. The panel's priority was to provide guidance on how best to further develop the market-oriented medical care components of the NIPAs (and the NHEAs); this strategy was in large part a reflection of our belief that these improvements are also essential building blocks for creating a national data system that extends and organizes information about the inputs and outputs of health. The panel was also asked to assess the extent to which existing data sources—such as those housed at CMS and the National Center for Health Statistics (NCHS)—and emerging sources—such as the new BLS American Time Use Survey—are meeting data needs for a coordinated health monitoring data system.

The audience for this report includes statistical agencies, research funding agencies, academic groups working on satellite health accounts, and Congress, which will be funding improvements to health data generally. For the most part, the audience categories parallel those into which the panel's recommendations fall. Our recommendations on constructing a medical care account are directed mainly to federal statistical agencies, including BEA and the agencies that provide source data to BEA, which include the Census Bureau, BLS, and NCHS. One extremely positive development is BEA's initiation of a project to build an account for medical care as a "satellite" account to the NIPAs. For this effort, we (1) assess what the agency could reasonably accomplish in the short term in light of what is currently known and what data are available, and (2) make recommendations to the source agencies that compile medical and health care data for their improvement and modification, with the aim of facilitating construction of the new medical care account.

The second audience includes researchers developing health data programs and the agencies funding these initiatives. For example, major topic areas such as measuring nonmarket and nonmedical inputs to health and defining and measuring population health are, for the foreseeable future, mainly the province of researchers and funding agencies interested in establishing links among medical care, health outcomes, and population well-being. For this group of recommendations, our guidance takes a tone of "here is what is still unknown and where more research and more (or different) data collection could fruitfully be commissioned."

Many (if not all) of the areas of emphasis by the statistical agencies—for example, medical care price indexes and expenditure accounts—are intermediate steps in the broader health data project. For example, an initial task for both is to get the spending-side categories and estimates in order and to confront the problem of how to distribute nominal expenditure estimates into meaningful production units to which prices and quantities can be attached. Indeed, this is really where both groups—those working on medical care accounts and on the linking of health inputs and outputs—are focused at the moment. Even where the projects do not currently overlap, they eventually may. For example, academic researchers are already well into thinking about how to track outcomes of the medical sector while statistical agencies are not. Ultimately, however, both BEA (NIPAs) and BLS (price index work) will also need to integrate medical outcome data because effectiveness of treatments (or preventive or diagnostic services), in terms of their impact on patient health, is the quality dimension that needs to be monitored in order to estimate changes in real prices and quantities correctly. As the statistical agencies continue work relevant to a satellite medical care account, they need to keep the broader and more ambitious ideas in mind so that the programs progress in a way that will not impede future work to track health care outcomes and quality change. Thus, those whose primary mission is funding research should be aware that the work under way at the statistical agencies, particularly BEA, and the data needed for this work are also part of the long-term research project to provide an accounting for the determinants of health.

The remainder of this report is structured as follows: Chapter 2 outlines the purpose and design elements of medical care and broader health accounts and discusses the conceptual bases for each. Boundaries for the medical care account are considered, and inputs and outputs are disentangled and catalogued. Chapter 3 focuses on the task of allocating medical care expenditures within a treatment-of-disease organizing framework. Disease classification schemes are assessed and the strengths and weaknesses of alternative approaches to allocating expenditures are identified. Some attention is given to data needs and alternative modeling approaches. Chapter 4 lays out strategies for measuring prices and quantities of medical care—essential for an economic account—through improved price indexes. Chapter 5 considers the role of population health measures in health accounting and offers guidance about which among the competing metrics are most promising. In Chapter 6, we discuss how data systems can be used to support disease modeling research oriented toward linking changes in population health to the array of inputs that affect it. Attention is given to how the U.S. health data infrastructure could be upgraded—in part through development of economic accounts—for this purpose.

The health account also implies a productivity measure, derived from equation 2.2:

$$\text{MFP change (health)} = \text{growth of health} / \text{growth of h} (\bullet) \qquad (2.4)$$

where h (•) is short for the right-hand side of equation 2.2. In this case, productivity change can be interpreted as a measure of efficiency in the use of all of society's resources that affect the population's health status. Computing MFP in the production of health requires, as it does in any MFP measurement, measuring all the inputs (or as many of them as can be identified) and finding an appropriate way to summarize them, analogous to the index number formula for inputs in equation 2.3. Representing the health function may be complicated because of the heterogeneity in the units in which the input variables are measured.

Partial productivity ratios may also be calculated. For example, measures of labor productivity are common. Based on equation 2.1, labor productivity (LP) growth in the medical care sector[9] can be expressed as:

$$\text{LP (medical care)} = \partial \text{ (medical services)} / \partial \text{ L}. \qquad (2.5)$$

Similarly, one may also be interested in the *productivity of the resources used in the medical care sector in the improvement in health*. Indeed, that is one of the most pressing policy issues of the day. Using equations 2.2 and 2.3, this measure of productivity growth (which we designate Ω) can be expressed as:

$$\Omega = \partial \text{ (health)} / \partial \text{ medical care} = \partial \text{ (health)} / \partial \text{ [f (KLEMS)]}, \qquad (2.6)$$

where ∂ designates partial derivatives of the variables in equation 2.2. As with any partial derivative, the value of Ω will depend on the values of the other variables in the equation. Thus, the productivity contribution of medical services to improved health will depend on the value of other health inputs such as diet and the environment.

Because the contribution of medical care to health depends on other health-determining factors, simple comparisons of changes in medical care and changes in health are seldom meaningful. That is

$$\Omega \neq \Delta \text{ (health)} / \Delta \text{ (medical care)}. \qquad (2.7)$$

One often hears such statements as the following: The U.S. medical care system must not be efficient or productive because the nation does not have the high-

[9]Triplett and Bosworth (2004, p. 263) reported that LP in the medical care services sector had negative growth from 1987 to 1995, but after 1995 it turned positive. They attributed the sign change to data improvements in the measure of hospital output in the second period, which demonstrates how measurement issues impact the analysis of medical care.

est health level in the world, even though it has the highest per capita spending on health care. This statement is fallacious logic because the impact of medical care expenditure on a nation's health cannot be established without considering nonmedical determinants of health. For example, benefits resulting from expenditures on cholesterol-reducing statin drugs may simply be offsetting greater than average obesity rates for the United States, so there may be no net gain in health relative to a society that is faring better in terms of related social determinants of health. Philipson and Posner (2008) suggest the inverse relationship between obesity and income is a consequence of a fall in the price of consuming calories and a rise in the cost of exercise (once a by-product of manual labor), which higher income groups can better afford: the medical system partly offsets the negative health impacts of obesity.[10] These examples demonstrate that the relationship between a nation's medical expenditures and its health is not a straightforward function of its per capita expenditures on medical care.

Measuring the nonmedical determinants of health, though essential, is very difficult, which is why assessing the impact of medical care on health is also very difficult. Separating the influences of medical and nonmedical determinants of health is why estimating an accounting for health would be so valuable.

2.2.2. Strategies for Going Forward

A solid case can be made for beginning with and emphasizing the medical care account. For policy purposes, the most pressing needs are to measure properly medical care expenditures and outputs, to improve measures of medical inflation, and to determine what part of increasing medical care costs are attributable to increases in medical services, as opposed to price change. In addition, accurate expenditure and output data on medical care, developed and presented by a detailed cost-of-disease metric, are essential for a "health" account. Finally, in terms of feasibility, much data on medical care goods and services already exist; the challenge is to improve these data and to array them using a more useful organizing principle.

In contrast, it will be difficult to collect data on the full range of behaviors and activities that affect population health. Nevertheless, nonmedical inputs to health matter greatly (McKeown, 1976; Mokyr, 1997); one cannot understand changes in health, and health differences between countries and population groups, without considering the nonmedical inputs. It is essential to begin to collect and maintain data on nonmedical and nonmarket inputs to health and to organize them into an analytic framework. Even if it is not currently possible to measure all the nonmedical determinants of health precisely, it is important to think through the measurement and conceptual issues.

---

[10]Michaud et al. (2009) and Lakdawalla, Goldman, and Shang (2005) provide empirical evidence that as much as 30 percent of the growth in spending on medical care in the United States can be linked to increased rates of obesity in the population.

2.2.3. Priorities: A Health Account or a Database on the Determinants of Health?

Grossman's (1972) "production function" approach to the analysis of health and its determinants (incorporated into equation 2.2, above) has become the standard for economists' thinking about the subject (see Bolin, Jacobson, and Lindgren, 2001). Yet attempts to implement the model empirically are few. The difficulties include the fact that health is multidimensional and—more problematic—that identifying and measuring its determinants are complex tasks.

Rosen and Cutler, in an ongoing project, are estimating disease models and combining them with economic data (see Chapter 3; Rosen and Cutler, 2007). Their research is couched within a Grossman-type framework and the related epidemiological perspective, so one can think of it as estimating equation 2.2 on a disease-by-disease basis. They have selected disease categories that account for a large portion of national health care expenditures and are working toward determining the factors—including medical care—that affect changes in mortality and morbidity. A second effort along similar lines for cancer and circulatory diseases in England is reported in Martin, Rice, and Smith (2008).

Beyond the Market: Designing Nonmarket Accounts for the United States (National Research Council, 2005) suggests constructing a health account that would provide a welfare-oriented measure as a counterpart to the market-oriented measures of the NIPA. Thus, it would be structured by analogy to the familiar national accounts that record economic activity but would be built around the functional relation and the variables in equation 2.2. *Beyond the Market* begins by identifying "gaps" in the existing national accounts that arise because their measures of outputs (and inputs) are incomplete—only market inputs and outputs are included. If the goal is to fill gaps in the NIPA coverage and structure, then it seems reasonable to work out an expanded accounting system that is patterned after the traditional national accounts structure.

However, assembling the data for an economic welfare account for health goes beyond the requirements for a database for research on health determinants. For example, *Beyond the Market* lists diet among health determinants, certainly an important consideration. The report of the World Cancer Research Fund/American Institute for Cancer Research (2007) summarizes evidence connecting dietary factors to different types of cancer—consumption of red and processed meats raises the risk of colorectal cancers, and excessive consumption of salt for stomach cancer, while consumption of fresh fruits and vegetables reduces risks of a number of digestive system cancers. A research model for cancers might be designed in which dietary data are employed in conjunction with medical care data to determine the relative impacts of diet and medical care on cancer death and incidence rates. If the objective is to estimate dietary determinants of health, then information on consumption of the foods of interest (data that are readily available) is the main requirement.

In the Grossman (1972) model, and from a welfare perspective, it is necessary to compute the *net* gain (Nordhaus, 2003), which is not the utility from improved health. It is the value of the increment to health, less the loss in utility from abstaining from steak and ham or from eating vegetables one does not like.[11] One can conceive of a research project to measure such utility losses and, though more difficult, perhaps include them in an aggregate welfare estimation. It is much less clear, however, how utility losses should be fitted into a welfare-oriented health account patterned on a NIPAs-type economic *accounting* structure.

One possible parallel is with environmental accounting. Some production processes (electricity generation, for example) produce both goods and "bads" (the bad in this case is pollution). In an environmental account, one subtracts the values of the bads from the goods to get a net welfare measure. In the health case, the bads are utility losses from pursuing more healthy lifestyles. The net output (the utility of health gains from changes in diet and lifestyle less the loss of utility from giving up things that give present utility but in the long run are deleterious to health) is the relevant measure for welfare and therefore for the NIPA-analog health account.

In principle, both objectives—an economic welfare account and a database for research on health determinants—should be pursued. However, in programs to generate data, choices must be made and priorities established. A database on the determinants of health that includes a measure of health has the most immediate policy value. It seems inevitable that estimating the utility loss from health-promoting lifestyle and dietary changes will compete with resources for moving forward on this work. If so, it is our judgment that the database for research on the determinants of health should receive higher priority. First, it is more immediately useful for a wide range of purposes. Second, information on the determinants of health is necessary for a welfare-type account in any case. Estimating the net gains from changes in lifestyles can follow.

Thus, it is not premature to recommend the collection of more information about the determinants of health. It may, however, be premature to recommend that statistical agencies organize health data to accommodate a health account of the welfare-oriented NIPAs type. As was true of the development of national economic accounts in the 1930s, health accounts have not yet evolved very far, but the situation should change as more work on their conceptual underpinnings and practical needs is undertaken.

The conflict between competing data needs is reduced the larger that the proportionate contribution of medical care is to the change in health. Historically, medical care was not the main determinant of past health improvements (McKeown, 1976). However, Cutler (2004) contends that medical care has been the main factor over the past half-century, particularly if one adds in the contribution of medical research to the information that has led to changed lifestyles. Even so, he concludes from an informal calculation that the loss of utility from

[11]On this point see also Philipson and Posner (2008).

of cleanliness that has medical implications and the part that augments custodial amenities. In general, the output vector in an industry account must line up with the inputs; accordingly, the boundary for an accounting of medical care outputs and inputs usually must be drawn along NAICS lines as a practical matter.[13]

In summary, in the medical care account, data constraints make it difficult to separate the strictly medical outputs of medical care industries from the value of nonmedical services that may be provided to patients as ancillaries to their treatments. It is clear that hospitals have extended their "hotel" functions over time, in response to patients' demands for improved living standards. These hotel-like amenities are provided more abundantly by U.S. hospitals than by hospitals in most other countries. At this point, it is not productive to spend much time debating the extent to which these ancillary services do or do not belong in a medical care account since, as a practical matter, it is unlikely that data could be found that would permit eliminating them from the existing totals on both the output and input sides.

The panel also considered other reasons for adopting a narrower definition. Some medical procedures have less direct or obvious health impacts. Examples that might be excluded, even though medical, are certain kinds of cosmetic surgery not covered by insurance (reconstructive surgeries are typically covered). However, the panel decided that trying to draw the boundary by recourse to insurance coverage was problematic, in part because insurance coverage lines are not uniform and are often in flux. For example, fertility treatments are sometimes covered and sometimes not; other categories of health care that are not uniformly covered by insurance include over-the-counter drugs and annual physicals. For many policy analytic and research purposes, it is important to track these kinds of treatments in the data system, even if they are not productive in improving population health status.

2.4. INPUTS TO THE MEDICAL CARE ACCOUNT

For the year 2002, BEA's industry account for NAICS 62, a major component of the ideal medical care account, indicated the breakdown of inputs to medical care shown in Table 2-2. Similar tabulations appear in the BEA industry accounts for NAICS subsectors 621, 622, and 623, described above. With the provisos that elements of household production should be included and that social services need to be removed from NAICS 62, as noted above, these data serve

[13]However, we note that if the output of custodial activities were removed from medical care as an input to the production of health, then the partial productivity measure discussed in equation 2.6, above, cannot be computed when industry data are used in the analysis, even though the MFP measure in equation 2.4 is defined. The reason is that medical care output, however defined by the researcher, enters into equation 2.4, but the resources used in medical care enter into equation 2.6, and if the resources are measured with industry data, the difficulties of separating inputs, discussed in the text, will apply.

TABLE 2-2 Inputs to the Medical Care Account ($ billions) (% of output)

Compensation	$553.8	48.7%
Taxes + imports – subsidies	8.4	0.7%
Gross operating surplus*	144.0	12.6%
Energy	11.9	1.0%
Purchased materials	119.0	10.5%
Purchased services	299.1	26.3%
Output	$1,136.2	

*This is (NAICS 62) System of National Accounts language for compensation of capital, essentially an expanded notion of profits and other returns to capital.
SOURCE: The estimates from the 2002 benchmark input-output (I-O) accounts at the summary level—135 commodities and 133 industries—and at the detailed level—428 commodities and 426 industries—are available on the Bureau of Economic Analysis website at http://www.bea.gov. Choose the "Industry" tab and then select "Interactive Tables: Input-Output." One can then select which variables to include in a customized table.

to introduce a discussion of data needs to support the input side of the medical care account. The discussion applies, unless otherwise stated, to inputs for the sector aggregate (NAICS 62 less social services) and to the inputs for the three individual subsectors recommended above. We focus on this industry account because of its central position in the medical care account.

Figures in Table 2-2 for NAICS sector 62 are the shares of inputs in the total receipts of the sector. Of course, expenditure on an input equals its price times the quantity of it that is purchased. Compensation of employees equals compensation per hour times the number of hours of employment in the subsector; the calculation is similar for other inputs such as purchased services. The change over time in medical sector expenditures on any input equals a period-to-period price change measure times the growth in the quantity of it purchased.

A full medical account would show each input represented by a price and a quantity change measure. For example, it would record not only total compensation, but also the hourly compensation (price of labor) and the amount of labor used as well; the same would be true for other inputs, including capital. The quantity change measures are essential for an analysis of the consumption of inputs, of productivity, and of the efficiency of the sector (as is true for the accounts for other sectors of the economy). The measurement problems associated with determining the shares of inputs to medical care are not particularly difficult (although there are some).[14] A much more challenging task is separating the changes in input shares into price and quantity elements. Improving price and quantity measures for medical care inputs provides the focus of the following sections.

[14]On determining shares, see Yuskavage (1996), who discusses difficulties in estimating the property income share and the intermediate input share.

2.4.1. Capital

Capital is the most complicated of the inputs to medical care, which justifies an extended discussion. Its contribution to output is the flow of services provided by the industry's stock of capital goods—physical capital (such as buildings and equipment) and also intangible capital (which includes, among other things, intellectual property).[15] The flow of capital services is derived from the stock (see Schreyer, Diewert, and Harrison, 2005). The price of capital services, in concept, is the charge for the use of a capital good for a unit of time. Estimating the price and quantity of capital services is complex because it requires (except when the capital good is rented or leased, and the lease payment provides the price measure) determining the capital stock, deflators for the stock, and measuring depreciation, all of which pose major empirical difficulties. However, medical capital goods pose no difficulties that are not already familiar ones in work on the topic for other sectors of the economy. The methods, accordingly, are not reviewed here (see Schreyer, 2001).

BEA now publishes capital stock for the medical care sector, tabulated in two ways:

1. by type: medical equipment and hospital and special care structures, and
2. by industry: an aggregate equipment and structures capital stock for all NAICS medical 3-digit subsectors.

The National Health Expenditure Accounts (NHEAs) have long contained data on investment in medical structures and now use BEA medical equipment data.[16] The NHEAs have traditionally also included investment in education in their accounts. However, the NHEAs do not estimate capital stock, and their flow-of-funds approach makes the estimation of capital services not relevant. Accordingly, the BEA capital estimates are the focus of our attention, beginning with a discussion of data needs for the capital services category.

The new BEA KLEMS input series contains the share of capital services in industry receipts for each industry, including data for NAICS 62 and its constituent subsector groupings. However, it is still true that BEA produces the capital stock series while the Bureau of Labor Statistics (BLS) estimates capital services. Even though the BEA stock data are inputs into the BLS services data, BEA does not include the quantity and price measures of capital services in its KLEMS

[15]A long history of debate over measurement concepts of capital in production analysis has spilled over into national accounts measurement. The debate is now settled on the lines suggested in Jorgenson, Gollop, and Fraumeni (1987). See, in particular, the Organisation for Economic Co-operation and Development handbook on productivity measurement by Schreyer (2001), which records the consensus and makes recommendations for the measurement of capital stocks and capital services that are wholly consistent with the methods now generally accepted in production analysis. From a conceptual view, medical care presents no unique problems in the measurement of capital.

[16]For a detailed discussion, see Sensenig and Donahoe (2006).

accounts (only their product). This imposes costs on researchers because related data must be obtained from different sources, with the attendant possibilities for errors in aligning the appropriate series, as well as creating the need for dealing with any inconsistencies among them. While BEA and BLS have historically worked well together, this is a case in which coordination could be improved.

Recommendation 2.2: The Bureau of Economic Analysis (BEA) should add capital services and capital services prices to the KLEMS measures in its medical care account. BEA should also make the measures available to data users on its website and in its publications.

Several relatively unpublicized yet vital data issues arise in obtaining capital measures for medical care industries. The problems are threefold.

First, for services industries the Census Bureau collects far less information on purchased inputs, including capital inputs, than it does for manufacturing. For example, on the 2007 Economic Census form for hospitals (NAICS 621), the only input information collected was employment and payroll, although a great amount of detail was collected for output activities and receipts. In contrast, the census form for the electromedical and electrotherapeutic equipment industry (NAICS 3345) lists, in addition to payroll, 12 classes of inputs, identified by Census Bureau material codes, for which expenditure information was collected.[17]

Hospitals, of course, are in services; electromedical equipment is in the manufacturing sector. Differences in their coverage of inputs stems from different Census Bureau policies for collecting information in the goods-producing and services-producing sectors. The resulting data incongruity handicaps analysis of services industries, including medical care industries, which require the same kinds of information that has long been provided for goods-producing sectors.

Critics of the U.S. medical care system have frequently asserted that it overuses imaging devices, partly because once this expensive equipment is put in place, the incremental cost of images is low, and they are often used for applications that have low value. It is accordingly bizarre from a research and policy analysis standpoint that data collections in the Economic Census do not even tell us how much imaging equipment is going into the medical care sector, let alone the total stock of it that is in place.

Recommendation 2.3: The Census Bureau should take steps to meet the urgent need for detailed data on inputs and capital goods purchases—especially those of technologically advanced intermediate inputs and capital goods—by service sectors such as medical care. Such information is important for research and policy analysis. If capital goods investment and stock informa-

[17]However, the data on output of medical equipment are not structured very well, a point we discuss in more detail below. In addition, none of the 12 input classes relates to purchased services, an input class that has been growing rapidly in all parts of the economy.

tion cannot be generated for service industries generally, then special surveys of capital goods purchases in the medical care sector should be undertaken, which should include not only investment but also stocks of technological equipment in place.

More detailed information supporting this recommendation appears in the annex to this chapter.

A second source of problems arises because surveys of industries that produce medical capital equipment have inadequate detail. Compounding the problem, the detail that is published suffers from inconsistent product classifications in various surveys that need to be used together and from archaic classifications that no longer match the technology. Evidence of the latter are surveys in which the largest product group is labeled "other" (which means that the detail is not useful). More information on this serious data problem appears in the annex to this chapter, which provides additional support for the following recommendation.

Recommendation 2.4: The Census Bureau, Bureau of Economic Analysis, and Bureau of Labor Statistics should work jointly to harmonize the classifications and structure of their published data on electromedical equipment and to reduce, so far as practical, the share of shipments falling into the "all other" categories in the Current Industrial Reports.

Deflators for equipment are the third major problem in medical capital data. All capital goods—including medical care capital goods—require price indexes as deflators, in order to convert information on expenditures into information on quantities.

BEA generally obtains capital goods deflators from the PPI and from the International Price Index (IPI) program. The PPI indexes for medical care capital equipment have serious shortcomings, and the IPI suffers from insufficient detail—even more so than the other measures we have discussed. The lack of agreed-on product lists, discussed above and in the annex, contributes to the inadequacy of the deflators.

PPI medical capital goods deflators include items falling within the following industry codes:

- 334510: electromedical apparatus manufacturing, which includes the following:
  334510-1: electromedical equipment, including
    medical therapy equipment,
    medical diagnostic equipment, and
    parts and accessories;
  334510-3: hearing aids;
- 339112: surgical and medical instrument manufacturing (including an index for catheters);

- 339113: surgical appliance and supplies (including an index for artificial joints);
- 339114: dental equipment and supplies (including separate indexes for equipment and supplies);
- 339115: ophthalmic goods (including separate indexes for plastic, glass, and contact lenses); and
- 334517: irradiation equipment manufacturing, which distinguishes ionizing equipment from all other types, but does not distinguish medical from nonmedical.

PPI industry codes parallel NAICS industries. Some, if not most, of the products in industries 33912-915 are intermediate products, and hearing aids (included in industry 334510) are not investment goods to the medical industries, although they should be included in medical care inputs.[18]

The medical equipment information presented in the PPI is quite sparse, particularly in the electromedical equipment area (334510). Aside from irradiation equipment, only two indexes exist for this area, and they are broad aggregates: medical therapeutic equipment and medical diagnostic equipment. Worse, these two cover only a part of the Current Industrial Reports (CIR) first-level aggregations, discussed in the annex, so they cannot be matched with Census Bureau data. As a result of both the scarcity of PPI indexes and the fact that they do not match Census Bureau data on shipments, BEA does not actually deflate medical equipment at the lowest level of PPI detail available, falling back instead to undesirable higher levels of aggregation.

Historically, the more difficult-to-measure commodities have had less PPI coverage, and medical equipment bears out this historical regularity. PPI medical equipment indexes number about the same as indexes for storage and primary batteries, which happen to lie just above medical equipment in the PPI commodity publication structure. Clearly, there is more inherent interest in, and need for analysis of, the medical care sector than for the production of storage batteries.

The PPI sample is drawn on a probability basis. The publication structure for the PPI largely depends on sample size. One solution to inadequate detail in PPI medical equipment indexes is for BLS to increase the sample size for medi-

[18]In addition, in the PPI product code structure (a different classification system from the industry structure), one finds PPI product code 11709-06, x-ray and electromedical equipment, which includes the following:

- 11790-0512 irradiation equipment;
- 11790-0514 diagnostic electromedical equipment;
- 11790-0516 electrotherapeutic equipment;
- 11790-0517 other electromedical equipment, excluding diagnostic and therapeutic; and
- 11790-0524 parts and accessories.

Indexes by product code do not always agree exactly with product code indexes from the industry code PPIs, for reasons that need not be explored here. We add this note because the users may find these indexes confusing.

cal equipment, in view of its importance for the analysis of medical care, and to increase the published detail for the PPI indexes in this category.

What should that detail be? Producing PPI indexes that do not match the Census Bureau's detailed data on shipments of medical equipment, or expenditures on them as investment goods, will bring little advantage. Therefore, expanding PPI detail should accompany and coordinate with the interagency task force suggested in Recommendation 2.4.

Turning now to the PPI indexes themselves, a number of questions can be posed about their present state of development. The first has to do with quality change of goods and services. For many years, economists have known that quality change poses serious price index measurement problems. Among the many relevant references that could be cited are Stigler (1961), Griliches (1964), and the Advisory Commission to Study the Consumer Price Index (1996). Quality change measurement errors are most likely when goods and services experience rapid technological change, and medical equipment has certainly experienced rapid technological change.

The two PPI electromedical equipment indexes are highly aggregated, as noted above. No doubt there are machines in them that have not experienced rapid technological change. Yet the relatively slow rates of decline of these indexes (a little less than 35 percent—total—over the past 20-plus years) seem low for categories that include such machines as scanners and imaging equipment.

A scanner is essentially an imaging device coupled to a computer. Vast strides have been made in imaging technology in recent years, and computers have declined in price at the rate of 20-30 percent per year for more than 50 years (see Triplett, 2003). Moreover, the only study of scanner prices that exists (Trajtenberg, 1989) found that CT scanner prices fell at a rate more like those of computers than the relatively modest (by computer standards) declines recorded in the PPI diagnostic equipment index. More accurate measurement of high-tech medical equipment is an urgent need.

Recommendation 2.5: The Bureau of Labor Statistics, the Bureau of Economic Analysis, and the industrial production division of the Federal Reserve Board should devote resources to addressing the quality change problem in price indexes for medical care equipment, using hedonic methods or other approaches, as the agencies see fit.[19] Agencies, including the National Science Foundation, that fund research on medical care issues should consider proposals for improving knowledge about the change in price of medical equipment and on methods for quantifying improvements in their technological capabilities.

---

[19]The IPI contains a category of computer and electronic equipment called "navigational, measuring, electromedical and control." Improvements to the electromedical measures might justify participation of the FRB in the efforts to improve data on medical equipment, which we would greatly encourage.

2.4.2. Labor

Intuitively, it seems that measuring the labor input—for which the basic unit is labor hours—should be easier than measuring capital. Yet data for the labor input to the medical care account contain serious shortcomings, some of which are not widely understood.

Human capital is a major contributor to output in all industries, but especially in medical care. A medical care account requires incorporating human capital adjustments to its measures of labor input. Otherwise, the account has the potential to seriously understate the amount of resources going into the production of medical care.

Methods for incorporating labor quality, or human capital, have been developed for use in industry accounts (Jorgenson, Gollop, and Fraumeni, 1987), and BLS includes a labor quality adjustment in its published MFP measures. These approaches are promising for the medical care account. Dawson and colleagues (2005) propose a human capital adjustment of this kind for the United Kingdom.

However, the traditional human capital measures distinguish mainly years of schooling or levels of educational attainment. They need to be extended in order to take into account the unique sets of skills and training in modern medicine. Although we do not know how this should be done, it is clearly an area for which research is needed. Indeed, the literature attempting to create human capital measures that portray the skilled labor input into medical care is extremely thin.

Recommendation 2.6: Research should be undertaken in statistical agencies, or funded by agencies interested in medical care topics, on the treatment of human capital in a medical care account. The Bureau of Economic Analysis should include human capital measures in the labor input for its medical care account (even at their present state of development).

In BEA's present industry accounts, nonmarket labor is not included in the labor input, because time spent in activities that are not paid is not included in GDP. However, volunteer labor in hospitals, hospices, and so forth and time spent caring for ill relatives and friends are important inputs in the production of medical care. In an experimental—or satellite—account, there is no reason to conform to all the conventions of national accounts, even if the medical care account retains the market output boundary of the NIPAs.[20]

[20]BEA defines satellite accounts as follows (Bureau of Economic Analysis, 1994, p. 41):

> [S]atellite accounts are frameworks designed to expand the analytical capacity of the economic accounts without overburdening them with detail or interfering with their general purpose orientation. Satellite accounts, which are meant to supplement, rather than replace, the existing accounts, organize information in an internally consistent way that suits the particular analytical focus at

BLS's American Time Use Survey offers relevant data. A researcher from BEA has already produced a paper on this topic, using these data (Christian, 2010). He found that time spent in home and volunteer production of health-related services was surprisingly low—about 14 percent of the total labor input to medical care. However, estimates generated from the National Health Interview Survey on disability suggest that the value of unpaid time providing long-term care services may be greater than the value of similar market-provided labor (LaPlante et al., 2002). This question needs review and additional research.

Recommendation 2.7: The Bureau of Economic Analysis should develop measures of nonmarket time devoted to the care of others and incorporate them into the labor input in its medical care account.

Even though employment itself may be easy to measure, that does not mean that U.S. measures of industry employment are robust. Triplett and Bosworth (2008), looking at three separate published measures of industry employment—two from BLS and one based on Census Bureau data—show that the three present surprisingly divergent information on industry employment trends. For all industry accounts, including those for medical care, divergences on the order found by Triplett and Bosworth suggest errors in existing labor productivity growth estimates as well as in other studies based on industry accounts.

Recommendation 2.8: The Bureau of Labor Statistics (BLS) and the Census Bureau should investigate and coordinate their surveys of employment by industry and take steps to eliminate or reconcile the substantial discrepancies that now appear in these data. Absent a full reconciliation, the Bureau of Economic Analysis should use the Census Bureau employment series in its medical care account, rather than the BLS employment series, because the census data are compatible with the other inputs in the account and with the output measure, all of which come from Census Bureau information.

> hand, while maintaining links to the existing accounts. In their most flexible application, they may use definitions and classifications that differ from those in the existing accounts. . . . In addition, satellite accounts typically add detail or other information, including nonmonetary information, about a particular aspect of the economy.

The United Nations System of National Accounts offers a similar description (United Nations, 1993b, pp. 45, 489):

> Satellite accounts provide a framework linked to the central accounts and which enables attention to be focused on a certain field or aspect of economic and social life in the context of national accounts; common examples are satellite accounts for the environment, or tourism, or unpaid household work. . . . Satellite accounts or systems generally stress the need to expand the analytical capacity of national accounting for selected areas of social concern in a flexible manner, without overburdening or disrupting the central system.

The report, *Understanding Business Dynamics* (National Research Council, 2007), provides a comprehensive set of recommendations for reconciling various business lists and data sources. It also discusses specifically how the Internal Revenue Service code needs revising to permit sharing data across statistical agencies to facilitate data improvements.

Another problem concerns the allocation of professional income between labor income and return to capital. This is an old issue in economics (see, for example, Friedman and Kuznets, 1945, and Christensen, 1969). Tax return data for partnerships and the self-employed, used extensively in the compilation of industry accounts, do not distinguish between the part of a professional's income that is properly a return to capital, as opposed to labor—what that professional could earn if employed by someone else. Because a large amount of medical services takes place in partnerships and individual practices, this issue looms larger for medical care industries than for some others, although it is indeed a problem for industry accounts generally.

The BLS productivity program has developed methods—applying ideas from Christensen (1969), Jorgenson, Gollop, and Fraumeni (1987), and elsewhere—for separating the capital and labor components. In the BLS application, the labor compensation of the self-employed is estimated from comparable employment in the same industry; the capital return is estimated from corporate entities in the same industry. Because the resulting estimates typically exceed the total income in the noncorporate sector, the labor and capital estimates are then compressed until they just exhaust the income total in the noncorporate sector. The BEA allocation is less robust (for more information, see Triplett and Bosworth, 2004). Going forward, BEA should, for the medical care account, do research to incorporate improved methods for separating labor income from property income.

2.4.3. Energy and Materials

Recent research on productivity and the analysis of production has accounted separately for energy inputs. BEA has already done so in its accounting for intermediate inputs in its industry accounts (Moyer, 2008), and there are no obvious special considerations with respect to the medical care sector's consumption of energy.

Material inputs to the medical care sector range over a wide variety of commodities, from writing paper and medical examination gowns to pharmaceuticals and stents. Moyer (2008) explains how BEA has recently improved its accounting for intermediate inputs, including purchased materials.

Some intermediate materials inputs to the medical sector present no unique problems. Hospitals consume paper products and cleaning supplies like other businesses. However, much recent technological change in medicine has proceeded through materials inputs that are unique to the medical care sector, including pharmaceuticals and medical devices such as stents. It is not clear how well these intermediates are represented in the current BEA medical account—the problems of coverage, of detail, of the quality of information on interindustry flows, and of matching Census Bureau receipts to PPI indexes that were discussed above in connection with high-tech capital inputs have parallels for high-tech intermediate materials. And again, problems of quality change in the measured deflators are paramount. Some good work on pharmaceuticals can be cited (see the summary in Berndt et al., 2000, and Danzon and Furukawa, 2008), but it is not nearly comprehensive. We are aware of no similar research on medical devices, despite their importance in some recent improvements in medical treatments.

Recommendation 2.9: For economic accounting purposes, work needs to be undertaken on the problems of accurately measuring prices, quantities, and the changing quality of medical devices and pharmaceuticals. Research funding agencies should consider these topics, which should also be on the research agendas of the statistical agencies.

2.4.4. Services

It has long been held that measuring the output of service-sector industries is difficult (Griliches, 1992, 1994). It is no less difficult when the services are inputs.

At the 3-digit NAICS level, purchased medical services include those from other medical industries, where they present problems similar to those of measuring medical outputs (section 2.5). In addition, purchased medical services contribute to the intraindustry or intrasector flows noted earlier. In particular, medical labs, imaging centers, blood banks, and so forth that are now (perhaps inappropriately) in the ambulatory care subsector will be removed from net output via the intervention of the intraindustry flows mechanism used by BLS and recommended above for the medical care account. But they are still inputs to the subsectors concerned, so they still represent measurement tasks. Most of them embody substantial technological elements so they pose similar quality change difficulties as those already discussed for medical capital equipment and for medical devices and pharmaceuticals. As was true of some other inputs, we know of little relevant research.

Successfully measuring the output of scanning and imaging centers (these centers provide imaging services to other medical industries) would substantially improve understanding of the role of imaging machines as capital goods. Looking at the outputs of various imaging machines might even be an alternative to direct investigation of the price movements of those machines. The PPI, however, contains an index for scanning, which does not differ greatly from the relevant index for electromedical equipment (see section 2.4.2, above).

2.4.5. Education

Medical education is treated as an investment in the NHEAs. In the usual framework for industry accounts, this investment in human capabilities is implicitly treated as a stock of human capital, the product of which is incorporated into the labor input. In other words, education augments the input of labor hours. Investment in education itself does not cumulate as a form of capital stock.[21]

Labor augmentation may not be the only way to think about medical education. For example, like most other forms of higher education, medical schools also produce research, and the production of research is integral to the training function. Because there is uncertainty about how to quantify any economies of scope in medical education, we support retaining for now the traditional treatment (medical education enhances the labor input in the production function for medical services). But we would also encourage imaginative research on alternatives.

2.4.6. Research and Development

Medical R&D poses questions that are similar to the discussion of medical education. Adding R&D to investment in national accounts has been proposed, and BEA has produced a satellite account that capitalizes R&D in the economy as a whole. If R&D is counted among capital assets, then the capital services provided by R&D would go into industry accounts as an input. Many issues surround methods for capitalizing R&D and for estimating capital services provided by R&D—and the literature on the topic is immense. The relevant materials for national accounts are summarized in Fraumeni and Okubo (2005).

Relative to other industries, medical R&D poses no unique issues. However, its treatment in a medical care account needs to be carefully thought through. Consider, for example, R&D that leads to development of an improved drug. If the effectiveness of the improved drug is accounted for in the medical care account (in which the drug is an input), as it should be, then also including the R&D that led to the improvement would double count. This is a kind of "stage of process" problem: pharmaceuticals are inputs to the medical care sector, so R&D on pharmaceuticals belongs in the industry account for pharmaceuticals, not in the industry account for medical care. R&D in the medical care account should include only R&D that is specific to medical care. In practice, however, it may be difficult to make the appropriate distinction.

[21]Another way to put it it that, for any form of capital, it is the services of the stock that enter into industry accounts. The services of the stock of education, being those of human capital, are entered into the industry accounts through labor quality augmentation.

Australian DRG system was derived from the U.S. one, and several European countries have developed similar DRG systems, Germany being a recent convert to the system.[23] Accordingly, economic data on a disease-based system, collected from providers, are increasingly available internationally, which adds to the practicality of measuring health care output on a disease-based metric.

The analytic advantages of a cost-of-disease system are its ultimate justification. Suppose, for example, that a heart attack patient is moved to a nursing facility for part of the recovery period in order to save higher hospital per-day costs; if so, hospital and total costs may fall, but nursing care costs rise. The traditional accounting (in the NHEAs) arrays spending by type of institution that receives the funds, so health policy analysts might see nursing home costs rising and conclude incorrectly that nursing care costs are a source of rising health care costs. Actually, rising nursing home costs may indicate cost saving, not cost increase.

In a cost-of-disease framework, the focus is on the total costs of circulatory disease in which all costs from whatever institution that receives payment are aggregated along disease lines. The total cost of treating a disease, not (or not mainly) how the cost is distributed among the industries/institutions, provides the appropriate focus.

As well, much other medical information, which an analyst would use in conjunction with medical care data and health accounts, is collected, tabulated, and organized on a similar principle. Scientific advances also fit into the system: research on new treatments takes place at the level of a specific disease, so it can be fitted naturally into the ICD. This is important for developing methods to deal with treatment improvements in output measures for medical care (discussed in Chapter 4). Thus, using a disease-based classification system such as the ICD for measuring medical care output makes the economic classifications line up with both the classifications that are used for other scientific work and the classification used for payments. This is a great advantage.

Cost of disease is the appropriate focus for consumer, insurance payer, and government policy perspectives. Essentially, all these are buyer perspectives, so the costs of the whole episode are what matter. Moreover, the analytic focus is on whether costs of, for example, circulatory or digestive diseases are growing more rapidly, not on the rate of increase in hospital costs compared with, say, nursing home costs.

2.5.2. From Concept to Practice

Cost-of-disease estimates require pricing the full series of elements in an episode of illness. The Organisation for Economic Co-operation and Development (2001) defines health care output as "the number of complete treatments with speci-

[23]However, Colecchia and Schreyer (2002) show that, of nine countries that have adopted DRG systems, each differed from the others.

fied bundles of characteristics" that "capture quality change and new products." A complete treatment is a "treatment pathway," in the language that has come into use, a measure that would encompass all contacts of a patient with the medical system, be they hospitals, doctors' or specialists' offices, clinics or other facilities. The Organisation for Economic Co-operation and Development's definition is consistent with much other recent writing of economists who have given attention to the subject. For example, Atkinson (2005, p. 113) writes: "Ideally, we should look at the whole course of treatment for an illness."

Going to a disease-based metric poses many practical difficulties, which we discuss in this and the following two chapters. To progress, major problems must be dealt with—arbitrariness around the definition of an episode, the issues of joint costs when multiple conditions are present (comorbidity) and treated (joint production—the situations where a single intervention improves more than one health condition), dealing with chronic episodes, and medical treatments (e.g., preventive) that do not fall neatly into disease categories. But these frequently voiced reservations need to be put into context: cost-of-disease accounts have been produced for many years, since Rice (1966). The challenge is not "Can it be done?" The challenge is to improve on what has been done in the past.

Modeling the Cost of Treating an Episode of Illness

The treatment pathway model takes its rationale from the patient's side. If it were also implemented with data from the patient, one would collect all the contacts, and all the bills, that are associated with one episode of illness—for example, a heart attack. One might collect all the claims paid by the patient's insurance and all the out-of-pocket expenses for medical care, plus any expenses that were incurred for which the provider was not compensated. Although sampling and data collection issues arise, the task seems straightforward.

The concept of a patient pathway inherently suggests data be collected from patients, as in the Medical Expenditure Panel Survey (MEPS), for example, or by aggregating insurance claims for a particular patient. However, an aggregation of patient pathways can also be implemented from the provider side.

In the 2007 Economic Census, hospitals (NAICS 622), offices of physicians (NAICS 6211-6213), outpatient care facilities (NAICS 6214), and diagnostic laboratories (NAICS 6215) were asked to provide receipts classified by ICD chapters (e.g., circulatory system diseases, respiratory system diseases, digestive system diseases). PPI indexes for hospitals have been constructed for many years using the same classification system, and BLS has proposed new PPI indexes for doctors' offices and medical labs that would mesh with the new Census Bureau data collections. These data are too new to have been tested and evaluated. However, they provide an option for constructing disease-based health care output measures that have not heretofore been available in the United States, nor in most other countries. With accurate measures for the different institutional units (subsectors) of medical care, the task is then to assemble these separate parts in a way that is motivated by the treatment pathway model.

Certainly, complications arise. For some illnesses, part of the cost is incurred in the hospital, part from fees paid to the doctor's office, part from laboratory charges, and part from nursing homes or other rehabilitation units. One might successfully collect from the hospital all the costs associated with its treatment of the patient, but the hospital would not usually know about services provided by other parts of the medical system or their costs. In principle, one could also collect costs from the patient's doctor and from labs and so forth that provided data on the patient's case. However, linking this information with hospital data is seldom easy in the absence of some system that provides patient linkages across providers (an insurer, for example, might have such data).

Two measurement problems arise for data collected from care providers: first, data must be linked across providers when only parts of the costs of an illness arise within any one provider. We contend this is not so serious a problem, despite what has sometimes been suggested. Second, treatments with the same outcome sometimes shift between providers in a way that lowers cost but that is not measured in traditional price index collections that draw repeated prices from the same seller. The latter problem, which is a serious one, is considered in Chapter 4 because (in the U.S. payments system) it is a price measurement problem.

Applying the Episode-of-Disease Model to Provider Data

The episode-of-disease model implies a patient's (or a purchaser's) perspective. To implement the model, it is natural to think of collecting data from patients or from insurance claims, from which one can in principle collect all the cost information at the same time or link costs from different sources via patient identifiers. Once the full cost of a disease has been collected at the patient (or claims) level, aggregate data on costs of diseases can be generated by aggregating individual cost-of-disease estimates over individuals.

The episode-of-disease model can also be applied to the provider side. Collecting cost-of-disease data from hospitals, doctors' offices, freestanding clinics, and so forth means, in effect, that *the aggregation over patients has already been done at the provider (industry) level.* Summing the cost data across industries then produces an aggregate cost-of-disease estimate that is, in principle, the same as that produced via patient or claims data.

To demonstrate, let c_{ij} be the cost of an episode of treatment for a specified disease for patient i at institution j (a hospital, say). Then $\Sigma_j c_{ij}$ is the total cost of the disease episode for patient i, aggregated across all the jth providers, as reported from individual or claims data. The aggregate cost of that disease is, aggregating across individuals, $\Sigma_i (\Sigma_j c_{ij})$.

For data collected from providers (via the Census Bureau–BLS mode), $\Sigma_i c_{ij}$ is the cost of disease collected from provider j (cumulating the costs of all the *ith* individuals who have interfaced with provider *j*).[24] Then, the aggregate cost of that disease is, aggregating across providers, $\Sigma_j (\Sigma_i c_{ij})$, which is the same aggregate as the one obtained from individual claims data.

It has sometimes been said that collecting from providers (as in the new Economic Census health industry data and the PPI price indexes) is deficient because the statistical agency does not collect the full spectrum of a patient's disease episode from any one provider. This is a misconception: aggregating cost-of-illness expenditures across providers gives the same aggregate cost-of-illness estimates, in principle, as aggregating cost-of-illness expenditures across patients.

Trade-Offs: Establishment Compared with Household Collections

As noted above, two choices exist for collecting cost-of-illness data. The data can be collected from individuals or insurers—from claims records, say. Alternatively, they can be collected from establishments that provide the services (the collection framework for Census Bureau collections of health care sector data and for BLS in the PPI).

Claims data offer much larger samples and facilitate linking costs for a disease episode across providers. Provider data offer the advantage of more precise control over the characteristics of the price and possibly greater accuracy of the totals.

Because cost-of-illness estimates are not fully developed, especially in time-series form, relatively little empirical information exists to compare the two alternatives. Thinking about the matter can benefit from considering nonmedical areas in which household and business (usually establishment) collections exist.

As one example, every month BLS publishes employment estimates from two surveys. One is a household survey, the Current Population Survey; the other is a survey of establishments, now called the Current Employment Survey (known informally as the 790 survey). The survey results are in fact reported in the same monthly BLS press release.

It is well known that the two surveys of employment yield differing estimates for monthly changes and also in some cases over longer time periods. The many "reconciliation" studies done over the years only partly account for the differences.[25] Although the dominant professional opinion puts more credence in employment estimates from the establishment survey, it is very hard to confirm the empirical basis for this belief. Household and establishment surveys of employment have differing and conflicting statistical properties, and those statistical properties and biases produce differences in employment estimates that have been hard to account for.

---

[24]The statistical agency will usually present this number as the *jth* agency's *receipts* from treating the disease.

[25]See the paper presented to the Federal Economic Statistics Advisory Committee, 2006, at http://www.bls.gov/bls/fesacp2120905.pdf.

As a second example, two estimates of national consumption also exist. BLS publishes the Consumer Expenditure Survey, which reports expenditures on consumption obtained from a household survey. BEA publishes personal consumption expenditures, a measure that is compiled mainly from business reports, such as retail trade. Again, an extensive literature exists (Triplett, 1997; National Research Council, 2002; Garner et al., 2006), and again the many attempted reconciliations have been only partly successful in explaining the differences. As with the employment example, professional opinion favors data from the business source, but in this case as well, the evidential basis for that belief is not so convincing as is the evidence that the two consumption data sources give different estimates.

Once medical economists have the luxury of comparing estimates from personal interview surveys, like MEPS, with business collections, such as the imminent Census-BLS medical care data, we anticipate that differences will provoke another reconciliation literature comparable to those that exist for employment and for consumption.[26] The pattern of those other cases suggests that personal interview surveys and other household side surveys will have great value, but that their advantages over data from business samples will not be overwhelming. At the present underdeveloped state of medical data, one cannot decide whether household or interview or claims-type medical data necessarily dominate medical cost data collected from establishments.

However, if the data collection also has the objective of generating microdata on patients, collecting information from the patient side (or from insurance company claims) is the only option. Microdata are useful for many research purposes.

Recommendation 2.10: The Bureau of Economic Analysis should construct cost-of-illness estimates for its medical care account and should consider the advantages and disadvantages of provider-side and patient-side (or claims) data for this purpose.

More detailed recommendations pertaining to this task appear in Chapter 3.

2.5.3. Measuring Spending on Nondisease-Specific Health Care Goods and Services

The treatment of diseases represents a large part of the activity of the health care sector. Roehrig and colleagues (2009) produced estimates of national health spending by medical condition using 260 categories defined in the Agency for Healthcare Research and Quality Clinical Classification Software, which groups

[26]There is already at least one published paper on MEPS-NHEA reconciliation—see Sing, Banthir, and Selden (2006). This study does not compare cost-of-illness results because the NHEA contains no information on cost of illness.

the numerous ICD-9 codes into broader categories that are "clinically meaningful."[27] In so doing, they provide support for the idea that accounting for major diseases covers a substantial portion of medical care activities in terms of costs.

The research team reallocated NHEAs totals using data from MEPS and a crosswalk methodology developed by Thorpe, Florence, and Joski (2004). While MEPS data are known to have some shortcomings (e.g., undercount of high-cost cases, which introduces a downward bias for some conditions), the estimates provide a general idea of the distribution of expenditures using one among many possible grouping structures. The circulatory system diagnostic category, which includes heart conditions (coronary heart disease, congestive heart failure, and dysrhythmias), as well as hypertension, cerebrovascular disease, and hyperlipidemia, accounted for the highest portion of costs, 17 percent of personal health spending in the United States in 2005. The next largest seven categories—ranging from mental disorders (9 percent) to nervous system disorders (6 percent)—accounted for about half of total expenditures. By contrast, prevention, exams, and dental—which would not fit cleanly into disease-specific categories—accounted for around 6 percent of personal health spending; another 6 percent of total personal health expenditures from the NHEAs was unallocated.

The Roehrig et al. estimates are in same ballpark as those produced in earlier studies. For example, Hodgson and Cohen (1999) found that the "big five" disease areas—(1) mental, (2) nervous system, (3) circulatory, (4) respiratory, and (5) digestive—accounted for about two-thirds of personal medical care expenditures.

These findings notwithstanding, medical care is clearly not limited to treatment of specific diseases. Long-term and preventive care are examples of services that do not always fall neatly into a disease classification. Some patients in long-term care are frail in many ways: their entry into a nursing home may be triggered by a disease—for example, disability due to a stroke— but their medical conditions are not tied to a single disease or health condition.

In addition, nursing homes provide rehabilitation and convalescent care in many episodes of disease. Ideally, one would divide nursing home days between episodes that are disease-specific and those that are not. Accordingly, the output of nursing homes is perhaps best measured as a day in a nursing home, with the days classified by level-of-care codes (such as those used for Medicare and Medicaid compensation). Because of the difficulty of otherwise separating nursing home and postdisease treatment (i.e., hospital "rehab" units) into the disease classification system, it may not be desirable to shoehorn all medical care spending (and associated health effects) into a disease-based classification system.

The same reservation applies to some instances of medical management—screening, diagnosis, and prevention. Although some are directly linked to a disease (mammography is clearly a preventive cost associated with cancer), many

[27]This research was presented to the panel at its March 14, 2008, workshop.

others are not. Yet quantities and prices of goods and services in these nondisease-specific categories of medical care should be tracked over time. When lab costs appear as inputs to the other medical sectors, they usually do so as costs of a specific ailment; but when lab output is a final product (tests done on behalf of the patient, for example), it is best measured in conventional ways. Prices and quantities of medical labs, in other words, are measured in terms of the tests the labs perform.

As well, doctor visits and many medical tests are sought for reassurance that a medical condition does not exist. Those contacts with the medical system contribute to well-being. Some are probably reported in claims records and so forth under the disease that is found not to be present, and there is a heading for "symptoms, signs, and ill-defined conditions" in the ICD.

Recommendation 2.11: Although starting with medical care on a disease-by-disease basis is a realistic way to proceed in order to begin accounting for a very significant share of the medical care economy, work should also begin on estimating the costs of, and eventually the health return from, interventions other than treating specific diseases (e.g., management, preventive, diagnostic, screening) and long-term medical services.

2.5.4. A Treatment Index or an Outcomes Index?

Equation 2.8 suggests that medical care interventions are valued by their incremental contributions to health—that is, the output of each intervention is its medical outcome measure. If so, why not measure medical outcomes directly, disease-by-disease, and combine them into a weighted measure, rather than forming measures of treatments? That is, why not ignore the intervention entirely and look only at its effect on health? Indeed, Dawson and colleagues (2005) proposed exactly that. Their preferred basic measure of the output of the National Health Service is a weighted index of quality-adjusted life years (QALYs), grouped by disease classifications, in which one QALY is valued at £30,000 (see their equations 12 and 111).[28]

Moreover, even a treatment-based system requires medical outcome measures. As spelled out in Chapter 4, medical outcomes are needed to adjust the output measure for improvements in treatments. The issue, then, is whether a treatment index should be constructed that is adjusted by medical outcomes, or whether an output index should be constructed that is composed entirely of medical outcomes, QALY or QALE (see Chapter 5), without recourse to counting or valuing treatments.

One reason for preferring the medical outcome measure is the general knowl-

---

[28]Garber and Phelps (1997) further substantiate the use of a QALY measure as an indicator of health care output.

edge that not all treatments are effective. Errors and mistakes, misprescriptions and misdiagnoses (patients still receive antibiotics for viral infections, for example), botched operations, and variance across areas in modes of treatment[29] are well known. Some interventions do not make a positive incremental contribution to health. For these cases, bypassing the treatment measure would bring the output measure closer to one that truly measures the incremental contribution to health that the medical care system makes.

Similar phenomena occur in other parts of the economy and are not adjusted out of national accounts. Botched and inappropriate car repairs, for example, occur with considerable frequency; sometimes they are corrected by the original repairer so the corrections do not result in new output, but sometimes the customer seeks out a new shop, so that repairing the botched job actually increases GDP. Such "redos" are not subtracted from GDP, even though they hardly contribute to consumers' welfare, nor is GDP adjusted for defective manufactured products that are also not infrequently produced.

But parallelism does not necessarily lead to good measurement practices. Methods for measuring medical output need to be considered on their own merits, apart from other national accounts practices, especially if the medical care account will be some form of alternative or satellite account, as seems likely.

Nevertheless, the distinction between output and welfare has a bearing on measurement principles. Particularly when there is to be a health account, in addition to a medical care account, adjusting medical care output is not the only way to handle defective and inappropriate treatments. Whether appropriate or not, treatments are still produced in the medical care sector, and they still use resources in the medical care sector. By that standard (the conventional way of looking at output), they are outputs of the medical care sector.

Determining whether or not medical sector output arising from inappropriate treatments contributes to welfare is a task for the compilers of the health account, particularly since they are more likely than national accountants to have the expertise to determine when treatments are not effective. When George Washington was bled as a treatment for pneumonia, his doctors must have thought they were contributing to his health, and the national accounts of his day, had they existed, would have recorded a treatment (or its resource use) in national output. When medical knowledge advanced enough to understand that Washington's treatment hastened his death, was the accounting revision that the advance in medical knowledge demanded best put into the national accounts measure of medical output or into the health account? The most important thing, surely, is that the revision be made. But for both consistency and expertise reasons, it seems better to make the revision in the health account— that is, national output for 1799

---

[29]Interarea differences in medical practices may be errors or may be differences of opinion about best practice. But even if the latter is true, presumably more knowledge will eventually show that some treatments that were thought to be best practice in some areas were in fact errors.

industries, which require the same kinds of information that has long been provided for goods-producing sectors. The same flaw carries through to the Annual Services Surveys, which also are deficient in input data.[31] The Census Bureau has instituted ACES to fill the gap, but this survey has been directed toward obtaining investment data for the economy as a whole, and its usefulness is greatly limited by inadequate industry and commodity detail.

Promising for our purposes is an information and communications technology (ICT) supplement to ACES that distinguishes the category "electromedical and electrotherapeutic equipment." This ICT category in ACES includes major types of equipment used in the medical care industries and matches the electromedical equipment category in the BEA capital flow table. The ACES survey form collects 4-digit North American Industry Classification System (NAICS) industry codes.

However, the promise of ACES has not been fulfilled. ACES published for 2005 and 2006 only at the 2-digit NAICS level (NAICS 62). No detail is published, only the total for investment in electromedical and electrotherapeutic equipment, plus information by type of acquisition (capitalized, leased, and so forth). ACES thus provides only a very limited benchmark—for one aggregated type of medical equipment, at the level of the medical sector as a whole.

Other relevant Census Bureau surveys collect data on U.S. production of medical equipment, not investment. The most informative, CIR, collects data on U.S. production of the products of NAICS 33451, electromedical apparatus manufacturing, and of some other medical equipment, at considerably more detail than ACES. ASM distinguish as the main product of NAICS 33451 "diagnostic and therapeutic" equipment, presumably the same "electromedical and electrotherapeutic equipment" products that are collected in ACES. No detail beyond this aggregate is published in ASM.[32]

The 2007 Economic Census form for the industry "Electromedical and Electrotherapeutic Apparatus" gathered information on receipts from "electromedical equipment including diagnostic, therapeutic and patient monitoring equipment." This is the same level of fairly gross aggregation as in ASM, and the Census Bureau form specifies that it is the same aggregate as on CIR. Thus, the Economic Census, which collects in many industries more detail than in annual collections, in this case does not approach the detail in the CIR.[33] The Economic Census also collected data for NAICS 33911; this industry makes nonelectronic medical equipment.

[31]This old data lacuna in services-producing industries is discussed more fully in Triplett and Bosworth (2004, Chapters 10 and 11).

[32]Both CIR and ASM record that production of these products was considerably greater than total U.S. investment in them in 2005-2006, suggesting that net exports were high, but it is known that, for some of these products, foreign producers are important suppliers.

[33]The Economic Census also collected information on other products, including irradiation equipment, scientific instruments, nonelectromedical surgical and medical apparatus, catheters, and so forth, that are also made in this industry (in which they are secondary products).

The problem of inadequate capital data is not unique to medical care industries. However, medical care is the largest sector of the economy for which detailed data are not available on purchased inputs, including capital inputs. Moreover, medical care industries purchase a range of very highly technological equipment, which is importantly linked with technical change. Thus, unlike some other services industries in which absence of capital expenditure detail is merely an annoyance (for example, NAICS 812, Personal and Laundry Services), in medical care the data gap threatens understanding of essential aspects of recent developments in the sector.

Critics of the U.S. medical care system have frequently asserted that it overuses imaging devices. It is accordingly bizarre from a research and policy analysis standpoint that data collections in the Economic Census do not reveal how much imaging equipment is going into the medical care sector, let alone the total stock of it that is in place.

Greater data detail on technological capital goods used by the medical care sector is essential. The model for improving medical equipment data is the data published for computers and office equipment—the second largest category of medical industry equipment investment and another notable category of technological investment products.

Some years ago, government data on computers and related equipment were as seriously undeveloped as medical equipment data are today. A multipronged effort by all three major statistical agencies (BEA, Census Bureau, Bureau of Labor Statistics [BLS]) involved

- a new and more relevantly descriptive system of product codes;
- improved and more detailed data on shipments and sales receipts by detailed product; and
- improved deflators that, using hedonic price index methods, allowed for the rapid rate of technological changes characteristic of nearly all electronic goods.

The value of this extensive data development exercise was shown in the analysis of the substantial influence of information technology investment in the post-1995 U.S. productivity expansion—see, for example, Jorgenson (2001) and Jorgenson and Stiroh (2000). Without the development of a comprehensive data set on the production of—and investment in—computer and related equipment, analysis of the role of information technology in the U.S. economy would have been, if not impossible, certainly greatly handicapped.[34]

Medical equipment performs a similar role in sparking, facilitating, and implementing technological innovation, except it does it exclusively in the medical care industries and not economy wide. Much anecdotal information exists

[34]The Federal Reserve Board has also contributed more recently to improving the deflators.

about medical equipment investment, but it is not quantified in the way data on other ICT equipment are for purposes of economic analysis. The lack of good information on medical equipment is one more way in which data for the analysis of medical care suffer from long-term neglect.

The first step in a data improvement project such as the one needed for medical equipment is getting agreement among the agencies on a common product classification scheme. This seemingly mundane task is necessary because otherwise data, especially from BLS and the Census Bureau but also from different Census Bureau surveys, do not fit together, and data expansions by individual programs and agencies proceed in inconsistent directions. Moreover, there is no center in the U.S. statistical system for coordinating such matters; it usually takes a special task force composed of agency representatives.[35] We explore ways for improving a specific type of capital equipment in greater detail below.

Part II: Product Detail for Electromedical Equipment Data

In this part, we discuss product detail for electromedical equipment and its inadequacy for producing consistent and meaningful data. Similar reviews could be carried out for the other categories of medical equipment and indeed for investment in medical structures, which have their own unique problems. The example we provide illustrates principles that we think should be followed.

In most goods-producing industries, the Census Bureau 10-digit commodity codes provide the standard for product nomenclature. In the case of electromedical equipment (primarily, NAICS 33451), the list is (the last four digits only are shown):

- 1100 electromedical equipment, including diagnostic, therapeutic, and patient monitoring equipment;
- 1103 magnetic resonance imaging equipment (MRI);
- 1106 ultrasound scanning devices;
- 1109 electrocardiograp;
- 1112 electroencephalograph and electromyograph;
- 1115 audiological equipment;
- 1118 endoscopic equipment;
- 1121 respiratory analysis equipment;
- 1124 all other medical diagnostic equipment; and
- 3100 electronic hearing aids.

The categories 1103-1124 are subdivisions of the first one. Hence, no distinction is made between uses. For example, ultrasound diagnostic equipment and ultrasound therapy equipment are in 1106.

---

[35]The industry classification system, NAICS, is an Office of Management and Budget standard, and there is an emerging NAPCS for products in the services sector. But no similar, formal standard exists for goods-sector products.

No one seems to use the Census Bureau 10-digit list for this industry. For example, the Census Bureau's CIR imposes use-categories as its first disaggregation:

- medical diagnostic equipment;
- patient monitoring equipment;
- medical therapy equipment;
- surgical systems; and
- other electromedical and electrotherapeutic apparatus.

CIR's second disaggregation is by product. But even though CIR presents much more product detail than the ASM or the Economic Census, the CIR published product detail does not map exactly into the Census Bureau 10-digit product list. For example, where do defibrillators go in the 10-digit list? They are not diagnostic, so there is not even a place for them in the "all other" grouping.

The classification used by BEA (in its investment series) more or less follows CIR's first-level disaggregation, even though BEA does not use CIR data for medical equipment investment. BEA's preferred classification scheme is also inconsistent with the Census Bureau 10-digit product codes.

Product codes in the PPI industry classification are also broadly consistent with CIR's first-level disaggregation, although a separate PPI commodity code scheme disaggregates differently. In addition to measures for the industry aggregate (electromedical apparatus), both the PPI and CIR contain data at the first-level disaggregation (that is, "diagnostic equipment" and so forth). But "medical diagnostic equipment" is still far too broad: CIR is right that meaningful disaggregation would produce series such as "ultrasound scanning devices" and "EKG."

The BEA and PPI codes suggest an incipient interagency agreement on the CIR first-level disaggregation, except for possibly the Economic Census and ASM. However, even incipient agreement between PPI and CIR does not yield detailed data on medical investment because BEA does not use the PPI for deflation at this level, and indeed it does not use CIR for any of its investment estimates. CIR contains much product detail, but the PPI is insufficiently fleshed out to match it.[36]

The CIR contains more detail, and more meaningful categories, than do any of the other surveys, including the PPI. At the product level, however, the CIR is problematic. For one thing, the size of CIR product categories labeled "all other" equipment makes the CIR categories less informative than they ought to be (see Table 2A-1). To take the worst case, the largest—by far—category of patient monitoring equipment is "all other patient monitoring" equipment; it accounts for 86 percent of the total. "All other" miscellaneous categories are not useful

[36]We leave aside any judgments about whether CIR data are equal in quality to ASM or Economic Census data.

3

Allocating Medical Expenditures: A Treatment-of-Disease Organizing Framework

3.1. MOTIVATION FOR DISEASE-BASED ACCOUNTS

Developing a national health account—whether the satellite medical care version of the Bureau of Economic Analysis (BEA) or a broader version designed to track population health status and its determinants—requires defining useful expenditure categories and then devising a method for allocating economy-wide spending on medical care into those categories. In addition, units of output that are meaningful from a consumer standpoint must be identified in such a way that price and quantity measures can be attached.[1]

In Chapter 2, we described the two existing accounts for medical care—(1) the National Income and Product Accounts (NIPAs) and (2) the National Health Expenditure Accounts (NHEAs)—and developed the profile of an improved and more adequate account that links medical care inputs with medical care output. Section 2.5 specifies the output concept for measuring the production of medical care: it is an episode of treatment for a disease. In the two existing accounts for medical care, however, the output concept is not fully developed, as neither account presents clear information on what the medical care system actually produces.

The NHEA have been compiled and maintained by the Office of the Actuary at the Centers for Medicare & Medicaid Services (CMS) since 1960. The accounts

[1] Health care purchasers are struggling with a distinct but not dissimilar challenge—determining how best to measure efficiency *within* the health care system (Leapfrog Group for Patient Safety and Bridges to Excellence, 2004; Pacific Business Group on Health, 2005; McGlynn, 2008; Physician Consortium for Performance Improvement® Work Group on Efficiency and Cost of Care, 2008). These efforts also require identifying meaningful (i.e., measurable and actionable) measures of health care output.

track the flow of funds into and out of the health care system, providing information on payer type (e.g., Medicare, out of pocket) and services purchased (e.g., hospital care, pharmaceuticals) in a series of standardized tables published annually on the CMS website. A typical NHEAs table forms a "sources and uses" matrix, imposing a specific set of accounting principles for who pays and how much, ensuring that all subtotals add up in a consistent manner.

While providing essential information on health care spending trends, the NHEAs have historically revealed little about the output of the sector—what is being bought—in terms that are meaningful for assessing medical care productivity and the impact on population health. The highly aggregated NHEAs data leave gaps that need filling if a number of critical health policy questions are to be resolved. Does an expensive new medical technology provide enough added health benefit to justify its use when compared with less costly alternatives? How do the public and private sectors encourage or limit adoption and diffusion of new technologies? And, more generally, which medical treatments are the most productive in terms of generating improved population health, and which are the least?

With the NHEAs alone, it is not possible to determine whether medical costs are increasing more because of cardiovascular disease treatments or because of cancer prevention activities. It is also largely unknown who is affected, and how, by the spending. Are vulnerable populations benefiting or suffering from current resource allocation strategies? Simply put, health care cost containment strategies in the United States are debated and pursued with inadequate information about what (or on whom) money is being spent (Triplett, 2001; Triplett and Bosworth, 2008). Addressing critical health policy questions requires more disaggregated data. Recognizing this need, there have been strong arguments for integrating cost-of-illness (COI) data into the NHEAs (Thorpe, 1999; Rosen and Cutler, 2007), linking microdata from national expenditure surveys to the macrodata in these accounts.

A similar deficiency limits the value of the medical care information in the NIPAs. Output estimates exist for the medical care sector and subsectors (for example, the ambulatory care subsector), but nowhere in the NIPAs is information presented on the products that the medical care sectors produce. Adding COI estimates to the NHEAs and the NIPAs can provide this critical information. Thus, a central issue in expanding either account of medical care is adding the disaggregated microdata needed to estimate treatment of disease costs.

As discussed in Chapter 2, linking health care spending to the treatment of specific diseases is useful in several respects. It provides a framework for understanding changes in the cost and quantity of health care, and it makes it possible to distinguish the effects of increasing prices for health care from the effects of increasing provision of services. Disease-based accounts also provide useful indicators of the economic burden individual diseases place on society; they can also be used to help identify how health resources are currently allocated, including across different population subgroups (informing questions of distributional equity). In addition, estimating health care expenditures by disease permits linkages with other essential information. For example, the effectiveness of therapies and the outcomes of care are measured this way, so a disease-based classification of spending is more clinically relevant and understandable to providers and to patients.

Of course, health care expenditures by themselves, even if grouped by disease, tell us little about health system performance or about priorities for resource allocation. Ultimately, if the links between spending on treatments and prevention (the inputs) can be successfully related to resultant changes in health status (outcomes), policy makers will be armed with a powerful tool—the information needed to better target spending to its most efficient uses. Ultimately, this tool will help determine who—the wealthy, the vulnerable, the elderly, the young—is benefiting or suffering from current resource allocation strategies.[2] Developing this information base is a complex, multistage process.

In shifting the output concept, we noted in Chapter 2 that an episode of treatment may not apply neatly to a category, such as preventive care, that is beyond those explicitly designated for diseases. Likewise, episodes may need to be specified differently for acute care than for chronic care of the same disease. In short, there are multiple ways in which episodes of care can be conceptualized, categorized, and put into practice for attributing spending across the range of medical services. In this chapter, we sort through some of these options and describe issues that must be resolved in order to move toward a treatment-of-disease framework for a national health account.[3]

3.2. COST-OF-ILLNESS ESTIMATION

While the NHEAs measure spending broadly by source and recipient, a separate literature has focused on measuring the costs of particular illnesses using more disaggregated data. These COI studies quantify the economic impact of a disease and, along with information on prevalence, morbidity, and mortality,

---

[2]COI estimates have been made for population subgroups. Moreover, the Bureau of Labor Statistics already controls for at least some demographic aspects when it prices diagnoses for the Producer Price Index. However, if such estimates are made by population subgroups, we might envision needing a price index for each, e.g., females with heart disease, Also, it is difficult to say whether there would be significant gains from adding demographic breakouts within the existing disease grouping detail—we do not know (very well) the variation in health status gains from various treatments across groups. Our intuition is to give priority to additional disease disaggregation over disaggregation by demographic group, but it is premature to answer, and perhaps even to consider, this empirical question now.

[3]Even if agreement is reached that episodes of care should be *the* unit of output, questions still exist about how to attribute the expenditures. Aggregation must take place at the person level if measures of health care output and health outcomes are to be comparable. It follows from our definition of quality that the unit for measuring medical sector output should be the patient treated. This makes it necessary to link the *activities* directed at treatment of a patient. For example, a patient undergoing treatment for heart disease would receive prescriptions for various drugs, attend outpatient clinics, and have lab tests. This topic is discussed in greater detail later in the chapter.

NHEA (Selden et al., 2001; Sing, Banthin, and Selden, 2006). Additional work by Rosen and Cutler (2007) has focused on linking MEPS and MCBS data in order to expand the scope of the covered population for reconciliation to NHEAs. It is encouraging that work to construct these data set linkages and reconciliations is progressing, but more research, with careful attention to detail, is required before they will be ready for use in an official statistical series. In the remainder of this section, we identify some of the more pressing challenges that will need to be addressed as reconciliation efforts mature.

Ensuring transparency about the scope of a national health account requires that decisions about which NHEAs cost categories to include be made explicit. Studies in some countries, such as the Netherlands, have often defined health care and costs of illness using a broad societal perspective (that may include "welfare" elements such as those related to informal care or the reduced well-being of family members due to morbidity and premature death) for their COI studies. U.S. studies, in contrast, including the research by Rosen and Cutler, have favored restricting the analysis to personal health care expenditures. There is no single right answer, but inclusion of nonpersonal health care does have one potential drawback when extended to health accounting: the method typically involves estimating the costs for a disease, not for persons with the disease. This implies that total costs for a disease can be translated to costs per capita but not so easily to costs per prevalent case of a disease. It also introduces types of spending and population groups that are most likely out-of-scope in the national surveys.

The implications of all the current and imminent data sets should be considered—for example, provider-side data would presumably pick up some excluded groups, but other coverage problems exist there as well. Researchers at the agencies working on this task will also confront the fact that, even for the covered populations, the scope of spending included in NHEAs and the surveys may differ. For example, NHEAs include total net revenues for all U.S. hospitals, but also government tax appropriations, nonpatient operating revenues (such as from gift shops), and nonoperating revenues (such as interest income) (Centers for Medicare & Medicaid Services, 2008). MEPS and MCBS, on the other hand, are event driven; most of these expenditures would not get picked up in the surveys, as they are not associated with discrete patient utilization events. Expenditures associated with discrete patient events (such as those going to freestanding labs and prescription medications) are underestimated as well. An approach implied by these data source characteristics is that provider-side data could be used as a control total, and the survey data on COI would then be used to help allocate across categories.

We discuss the characteristics and coverage of these national surveys in Chapter 6 in more detail because both expenditure- and outcomes-side data are needed. The remainder of this chapter focuses primarily on attributing expenditures to diseases assuming that national survey data will be the primary source of person-level estimates. We comment on some of the resultant challenges, leaving detailed discussion of data challenges and needs to the final chapter of the report.

3.5. DISEASE CLASSIFICATION SCHEMA

A wide variety of disease classification schemas exist, all differing with respect to the requisite data elements, populations covered, units of analysis, time period to which the assessment is applicable, and, at the most basic level, the types of analyses each is designed to support. For example, the cost category for diabetes could separate or combine type I and type II cases. Furthermore, patients with complications could be differentiated from those without, or everyone could be left in one spending category. There is no obvious rule about which strategy is best. Most systems use the International Classification of Diseases (ICD), 9th/10th revision codes; however, the number of disease categories and the combination of codes mapping to a given disease can vary significantly across systems, pointing to a need for comparative research. Furthermore, it appears that many systems start with the ICD chapters or with some other existing classification schema and then add or subtract categories to adapt to local conditions, such as clinical practice. While this may help tailor the classification system to users' needs, it makes standardization efforts difficult.

The validity of disease classifications can be optimized, in part, by grouping diagnoses into homogenous, mutually exclusive, exhaustive categories. However, the first-level categorization of the International Classification of Diseases-Ninth Revision, Clinical Modification (ICD-9-CM) (the most frequently used system in the United States) violates this rule, as do even the most detailed system entries. ICD-9-CM codes are organized into 17 broad categories or chapters—some represent organ systems (e.g., circulatory diseases, respiratory diseases); others represent conditions that span multiple organ systems (e.g., infectious and parasitic diseases, neoplasms); and one additional category is reserved for "symptoms, signs, and ill-defined conditions." As a result, for many purposes, the chapters range from too narrow to too broad. Recognizing that the chapters, or an appropriate combination of chapters and subchapters, make up the schema for publication does not imply that they are adequate for grouping observations, which in principle should be at a much lower level. A related problem is that two different, not fully compatible versions of the ICD are in common use (ICD-9 and ICD-10).

One categorization schema, AHRQ's Clinical Classification Software (CCS) (Elixhauser, Steiner, and Palmer, 2006), is unique in that it groups diseases with similar etiologies together, regardless of whether they cross organ system (or ICD-9 chapter) boundaries. This consistency, along with AHRQ's ongoing and timely maintenance of the CCS (updated annually to capture the frequent changes to ICD-9 codes), makes it an appealing instrument for standardization efforts. At the same time, though, the inconsistencies in many of the other grouping systems have made mapping to CCS challenging as well (Lu and Tsiatis, 2005).

A variety of commercial risk adjustment tools—such as Medical Episode Groups, Episode Treatment Groups, and Diagnosis Cost Groups—have also been used as the basis for disease categorization schemas. While, to our knowledge, no comprehensive catalogue exists, there have been two excellent recent reviews of many of these disease classification systems, one developed for clinical outcomes (Lu and Tsiatis, 2005) and the other for risk-adjusting costs (Winkelman and Mehmud, 2007). In the first study, seven grouping schemes—five for mortality and two for morbidity—were evaluated, and poor comparability was found to exist between them. The various schemas used different grouping logic, covered different ranges of codes, and named some groups the same but defined them with entirely different diagnostic codes. It is noteworthy that this set of divergent grouping schemes are the ones used to make most international mortality comparisons (Lu and Tsiatis, 2005). The second review, by the Society of Actuaries, made side-by-side comparisons of 12 largely commercial claims-based, risk-adjustment models. The models varied markedly in the data fields used to define patient risk categories and their output. For example, risk-adjustment tools may or may not include age, sex, and secondary diagnoses. Some included pharmacy and laboratory data, while others did not. The number of risk categories varied substantially, as did the proportion of expenditures that could be allocated to disease groups (Winkelman and Mehmud, 2007).

Problems associated with this kind of modeling might largely be solved through adoption of a standardized list of diseases, which would make it possible to map local classifications onto the list. Unfortunately, such a list does not yet exist, although some progress has been made. The World Health Organization has, in collaboration with OECD, Eurostat, and the Nordic Medico-Statistical Committee, recently developed an international short list for the tabulation of hospital data (World Health Organization, 2009). This is a useful point of departure for discussion, but the disadvantage for use in a COI analysis is that it was developed specifically for use with hospital data and may not be well suited for use with other providers.

In identifying an appropriate disease classification system, the number of categories will depend on the available data and current scientific knowledge. If this number is large, a two-level classification can be developed with a more aggregated level analogous to ICD-9 chapters and a disease level within these chapters (Slobbe, Heijink, and Polder, 2007; Heijink et al., 2008; Organisation for Economic Co-operation and Development, 2009). For each chapter, key diseases can be broken out (e.g., diabetes) with others mapped to a residual or "other" group (e.g., other endocrine diseases). When attributing expenditures to diseases, two additional categories will be needed as well: "disease unknown" and "not disease-related" (Slobbe, Heijink, and Polder, 2007).

The key point here is that, given the number of different disease classification schemas currently in use in the U.S. health care system, it is essential for all players, in both public and private sectors, to participate in and come to a consensus on the development of a single unified version.

Recommendation 3.1: A concerted effort is needed to reach consensus on how to classify diseases and about what the criteria are by which diseases are disaggregated from the very broad International Classification of Diseases chapters. The National Center for Health Statistics should lead the effort, working with the Agency for Healthcare Research and Quality, the Centers for Medicare & Medicaid Services, the Bureau of Economic Analysis, and other relevant statistical agencies. As part of this effort, U.S. agencies should participate in ongoing standardization efforts (such as those sponsored by the Organisation for Economic Co-operation and Development or the World Health Organization) to benefit from international expertise, to consider these as the basis for a national system, and to facilitate international comparisons.

The basic principles underlying the groupings are that they should be clinically meaningful, derived from routinely collected data (to the extent possible), and limited to a manageable number of categories. The criteria must ensure practicality as well as acceptability by the medical and economic communities.

Conceptually, the idea of having to choose groupings of ICD chapters and subchapters is not a major hurdle. The ICD already has a rich disaggregation below the hierarchy; however, resources are needed to sample entries for detail below the chapter headings.

Once a common disease classification system is chosen, the next step will be to implement it in order to generate data that are useful for medical care and health accounting purposes.

Recommendation 3.2: Using a population subsample for which good data exist, a pilot study should be undertaken by the Bureau of Economic Analysis using a proposed classification system with the goal of identifying adaptations needed for a national health account. At the point when the classification schema has undergone initial rounds of revisions and modifications, a concerted effort should be undertaken to consistently measure these diseases in the national health surveys in order to more accurately capture their epidemiologies.

Implementing these recommendations will create a foundation from which more targeted research can be conducted to attribute spending to the diseases. This step can be separated into the aggregation task, which can be thought of as the unit of analysis, and attribution, which is the method by which spending gets assigned to specific diseases. Because the two concepts are largely inseparable—the attribution method will almost always determine the level of

length of the "clean periods" that signal the end of an episode and the beginning of another.

To date, these episode groupers have not been adequately vetted by research (McGlynn, 2008). While a number of alternate episode groupers are already widely in use, they have received little scientific evaluation to date, and they have not been extensively tested for reliability, validity, or agreement with each other (McGlynn, 2008). A small but growing body of research by CMS (MaCurdy et al., 2008) and others points to significant variation in the output of different vendors' groupers. Perhaps most problematic is that the episode-grouping algorithms are proprietary and largely a black box, making it difficult to use them for public work.

Beyond the grouper-specific issues, comorbidities and the resulting joint costs are major challenges with the episode-based approach as well. It is common for individuals to receive treatment for multiple diseases simultaneously, and these comorbidities can lead to a very complicated picture of episode definitions and measurement (Hornbrook et al., 1985). Even in the absence of comorbidities, other challenges arise. It is often difficult (or not possible) to link data when the episode's services are supplied by several different providers. For chronic disease episodes, length of the episode must be determined (it is often set arbitrarily at one year). Complications of treatment for one condition may lead to the development of another. Should these be treated as a new episode or an old one? And, as we have pointed out, medical treatments do not always fall neatly into a disease category.

3.6.3. Person- (or Population-)Based Approach

A conceptually similar, but alternative, way of attributing expenditures on medical care to disease categories is to use a person-based approach. The distinction here is that spending is assigned to the entire set of diseases a person has, not simply to the primary diagnosis listed on a claim. Indeed, person-based (or population-based) measures were the norm before the introduction and rapid gain in use of the proprietary episode groupers. The key feature of these case-mix measures is that individuals—rather than episodes—are classified into clinical categories based on similar demographic and clinical characteristics (and grouper software exists for this purpose as well). Again, the goal is to categorize patients into relatively homogenous groups with respect to resource needs over some specified time window (usually 1 year).

In this approach, an individual's total health care spending over the period is regressed on indicators for the presence of all medical conditions. This approach is designed to produce more valid estimates for patients with multiple chronic conditions, as it better captures expenditures for comorbidities and complications. That said, the regression specification typically assumes that comorbidities have an independent effect on spending with few interaction terms included in the models. An empirical issue is what interaction terms to include. For the most part, clinical expertise is needed to identify the appropriate groups of co-occurring diseases; while clinical insight is likely to result in better estimates, the need for clinical expertise represents a limitation as well, particularly for federal agencies that have not typically had access to clinician health services researchers. There is also the issue of how to allocate the intercepts for the base spending for the year.

These criticisms aside, an attractive conceptual feature of person-based cost estimates is that they can be readily matched to health outcomes (a topic of Chapter 6), such as mortality and quality of life, thereby providing the critical link between spending and health needed to measure value more systematically. Furthermore, unlike the encounter- and episode-costing approaches, a person approach can conceivably attribute the costs of cases for which there are no valid claims or ICD-9 codes.

The person-based measures have been more thoroughly studied than have the episode-based measures, largely because they are older and many were developed by health services researchers. Many of the measures are statistically valid, are easy to implement, and have good predictive ability for explaining variation in utilization (Ellis et al., 1996; Weiner et al., 1996; Rosen, 2001; Iezzoni, 2003) when used in the populations for which they were developed (Arbitman, 1986; Hornbrook et al., 1991; Rosen, 2001; Iezzoni, 2003).

Person-based measures have limitations as well. First, the different groupers vary markedly in the data inputs required to define patient risk categories and then, in turn, outputs. For example, the groupers may or may not include age, gender, or secondary diagnoses. Some include laboratory, pharmacy, and/or procedure data, while others do not (Rosen, 2001; Zhao et al., 2001; Grazier, Thomas, and Ward, 2002, 2006; Iezzoni, 2003). Second, the effective sample size is smaller with person-based than with episode-based measures (e.g., one individual can have multiple episodes in a year). Because of this, while more patient groups are desirable to increase the homogeneity of expected resource use within groups, this must be balanced against the smaller sample sizes.

3.6.4. Comparing Episode- and Person-Based Methods

Although the evidence base is limited, researchers have begun comparing the different measures for attributing costs to illnesses. Thomas, Grazier, and Ward (2004) tested the consistency of six groupers (some episode-based and some person-based) for measuring the costs of primary care providers and found moderate to high agreement (weighted kappa = 0.51 to 0.73) between physician efficiency rankings using the different measures. In contrast, the Medicare Payment Advisory Commission (2006) compared episode-and person-based measures in area-level analyses and found they can produce different results. For example, compared with Minneapolis, Miami had lower average costs per coronary artery disease episode but higher average per capita costs due to a higher volume of episodes. Box 3-1

BOX 3-1
Examples of Episode- and Person-Based Case-Mix Measures

Episode-Based Measures

Episode Treatment Groups (ETG). The ETG methodology identifies and classifies episodes of care, defined as unique occurrences of clinical conditions for individuals and the services involved in diagnosing, managing, and treating that condition. Using inpatient, ambulatory care, and pharmaceutical claims, the ETG classification system groups diagnosis, procedure, and pharmacy (National Drug Code) codes into 574 clinically homogenous groups, which can serve as analytic units for assessing and benchmarking health care utilization, demand, and management. Developer: IHCIS-Symmetry of Ingenix.

Medstat Episode Groups (MEG). MEG is an episode-of-care measurement tool predicated on clinical definition of illness severity. Disease stage is driven by the natural history and progression of the disease and not by the treatments involved. Based on the disease-staging classification system, inpatient, outpatient, and pharmacy claims are clustered into approximately 550 clinically homogenous disease categories. Clustering logic (i.e., episode construction) includes (1) starting points, (2) episode duration, (3) multiple diagnosis codes, (4) look-back mechanism, (5) inclusion of nonspecific coding, and (6) drug claims. Developer: Thomson Medstat.

Cave Grouper. The CCGroup Marketbasket System compares physician efficiency and effectiveness to a specialty-specific peer group using a standardized set of prevalent medical condition episodes with the intent of minimizing the influence of patient case-mix (or health status) differences and methodology statistical errors. The Cave Grouper groups over 14,000 unique ICD-9 diagnosis codes into 526 meaningful medical conditions. The CCGroup Efficiency Care Module takes the output from the Cave Grouper and develops specialty-specific physician efficiency scores that compare individual physician efficiency (or physician group efficiency) against the efficiency of a peer group of interest. Developer: Cave Consulting Group.

Person-Based Measures

Relative Resource Use (RRU). RRUs report the average RRU for health plan members with a particular condition compared with their risk-adjusted peers. Standardized prices are used to focus on the quantities of resources used. Quality measures for the same conditions are reported concurrently. Developer: National Committee for Quality Assurance.

Adjusted Clinical Groups (ACGs). ACGs group illnesses into morbidity clusters rather than specific diseases. The ACGs algorithm assigns individuals into one of 93 mutually exclusive groups based on the individual's age, gender, and diagnosis codes (e.g., comorbidities). Clustering is based on (1) duration of condition, (2) severity of condition, (3) diagnostic certainty, (4) etiology of condition, and (5) specialty care involvement. Developer: Johns Hopkins University.

Clinical Risk Grouping (CRG). The CRG methodology generates hierarchical, mutually exclusive risk groups using administrative claims data, diagnosis codes, and procedure codes. At the foundation of this classification system are 269 base CRGs, which can be further categorized according to levels of illness severity. Clustering logic is based on (1) the nature and extent of an individual's underlying chronic illness, (2) a combination of chronic conditions involving multiple organ systems, and (3) further refined by specification of severity of illness in each category. Developer: 3M Health Information Systems.

Diagnostic Cost Groups (DCG) and RxGroups. DCGs begin with 118 condition categories determined by age, gender, and diagnosis codes; RxGroups add pharmacy data as an input. Both models create coherent clinical groupings and employ hierarchies and interactions to create a summary measure, the "relative risk score," which can be used to predict health care utilization. At the highest level of the classification system are 30 aggregated condition categories, which are subclassified into 118 condition categories organized by organ system or disease group. Developer: DxCG.

Provider Performance Measurement System (PPMS). The PPMS examines the systematic effects of health services resources that a person, at a given level of comorbidity, uses over a predetermined period of time (usually one year). The measures incorporate both facility/setting (e.g., use of emergency department and inpatient services) and types of professional services provided (e.g., physician services, imaging studies, laboratory services). Based on John Wennberg's work, PPMS assesses and attributes unwarranted variations in the system with respect to three dimensions: (1) effective care, (2) preference-sensitive care, and (3) supply-sensitive care. Developer: Health Dialog.

SOURCE: Adapted from McGlynn (2008, Table 7).

provides a more detailed description of several commercial episode- and person-based case-mix measures (McGlynn, 2008).[8]

Over the past year or two, the Cutler-Rosen group has been working with BEA to empirically assess quantitative differences in the various approaches to allocating medical care expenditures by disease; Rosen reported some preliminary findings at the panel's workshop. The research objective is to reconcile disease categories among the encounter-, episode-, and person-based regression approaches; to simulate costs of diseases using each; and to compare and contrast the findings. For this project, health claims data from Pharmetrics, Inc., for the period 2003-2005 were used. For 2003, the data cover just over 3 million lives and include total spending of $9.09 billion on inpatient and outpatient services, office visits, prescription drugs, skilled nursing facilities, and laboratory services. Up to four ICD-9 diagnoses are present on a given claim, although only the primary diagnosis is listed for hospital claims. Symmetry software from Ingenix was used to link medical expenditures to disease categories.

In order to reconcile the three approaches to common disease categories, the researchers first mapped ICD-9 codes into CCS categories. These were aggregated into 65 clinically meaningful groups that had been developed earlier based on clinical advice from physicians. Cost categories were created primarily for diseases with known treatments that have led to health benefits and for which more detailed analyses could be done matching quality to costs.

For the person-based regression approach, the authors were able to use all listed diagnoses on claims in a given year. For the encounter approach, an algorithm was used to determine the diagnosis (usually the first listed) to which the majority of spending went, and the dollars were assigned to that category. For the episode-based approach, each episode treatment group (ETG) was allocated to the clinical group that accounted for the largest share of spending. There were a few problems with the ETG approach. For example, there was no ETG for cervical cancer; also, the transparency issue remained—the method of aggregating data into the ETG was still essentially a black box.

Under the encounter approach, about 19 percent of the spending recorded from claims had no listed diagnosis, and the dollars could not be allocated to a condition.[9] Using the episode-based approach, only 1 percent of spending originated from claims with no ETG. Using the person-based approach, expenditures for individuals with no diagnoses accounted for only about 0.6 percent

[8]The other potential unit of output, the visit or encounter (discussed earlier), also uses groupers (e.g., case-mix measure) to classify the resources utilized. As with the other measures, the spending sorted into each visit-based group has similar diagnostic codes, and the hope is that the groups would have similar expected resource use and cost. An example is Ambulatory Patient Groups, which was developed by 3M (Averill et al., 1990). The system was designed to explain the amount and types of resources used in Medicare hospital-based outpatient visits. A shortcoming is its restriction to outpatient care.

[9]Hodgson and Cohen (1999), using an encounter approach, reported only 10 percent of expenditures with unallocated diagnoses. It would be enlightening to compare these results.

TABLE 3-1 Annual Per-Patient Cost for Selected Diseases by Method, 2003

Disease	Encounter	Episode	Person
Colon cancer	$8,100	$4,458	$10,475
Lung cancer	12,082	14,213	23,895
Dementia	596	1,111	9,231
Depression and bipolar disease	616	984	1,070
Hypertension	225	522	376
Coronary atherosclerosis disease	3,415	4,342	3,303
Congestive heart failure	2,869	2,476	12,645
Cerebrovascular disease	2,563	2,818	5,759
Asthma	348	639	519
Chronic renal failure	11,105	11,433	11,964
Osteoarthritis	1,184	1,726	1,450

SOURCE: National Research Council (2009, Table 2-4).

of the total. The problem of unlinkable spending was clearly most serious with the encounter-based approach.

The Cutler-Rosen work demonstrates that the cost of illness can be estimated by each of the proposed methods. The total dollar amount that can be allocated differs and, certainly, the fact that noncomparable data sets are being used for the different methods also has an impact on the results. Table 3-1 shows annual spending by condition estimates. The spending estimates varied significantly for some disease categories. For example, the person-based approach yielded very high annual expenditures for dementia—on the order of $9,000—relative to the encounter- and episode-based approaches. The likely reason is that the regression used does not include all of the needed interaction terms, so the estimates essentially capture unobserved correlates of spending. Instead of getting just spending on dementia, the coefficient is picking up aspiration pneumonias, feeding tube treatments, and all sorts of other things for which clinically meaningful categories need to be created. This illustrates why it is important to bring clinical insight into analyses.

For some of the same disease categories, the encounter-based approach appears to underestimate expenditures. This may have something to do with the way risk factors for diseases are commonly coded. For example, physicians may be more likely to code coronary heart disease than they are diabetes, hypertension, or hyperlipidemia. In contrast to the encounter-based approach, which relies entirely on physician coding on claims, one useful feature of the person-based approach is that coding can be captured over time, so more information about multiple conditions can be obtained; the approach also allows the claims data to be supplemented with surveys, injecting information from patients that can enrich the picture.

The research team has not yet done any time series with MEPS. Making some direct comparisons, they have found that some of the same things—for example, dementia or acute renal failure, which tend to occur in patients being treated for other conditions simultaneously—end up being much higher with the person-based regression approach than the others.

Ultimately, the choice of episode-based versus population-based measures will depend on the context in which they are to be used (Luft, 2006; Davis, 2007; McGlynn, 2008; Miller, 2008, 2009a; Mechanic and Altman, 2009). For example, while care of acute conditions may best be understood at the episode level, chronic disease care (and the provision of preventive services, such as cancer screening) may be better understood at the person level. For a given setting, the predominant provider payment approach may also impact the choice of measure. Whereas fee-for-service payments make episodes somewhat easier to interpret, capitation could be more readily evaluated at the person level.

So, which approach is best? At this point, the panel cannot definitively endorse one method for allocating expenditures to diseases over others. Rather, the best method depends largely on the question at hand and the needs of the target audience. For example, if the goal is to compare costs and health effects for a given disease, as is done in cost-effectiveness analyses, a person-based approach is likely to be most appropriate. In contrast, if price index construction is the goal, federal agencies may find an episode-of-treatment approach more meaningful. For a manager of a health plan trying to understand why emergency room spending patterns have changed, real-time answers may be possible only with an encounter-based approach. The choice of method will also invariably be constrained by the availability of data.

In the long term, what is needed is more empirical work to compare different approaches and to determine more definitively which is best under different conditions. BEA, for its part, is working both internally and with the Cutler-Rosen team to establish what the allocations end up looking like under the different methods and whether it matters for estimating expenditures and prices. These researchers have already begun producing disease-based cost estimates; spending could also be further broken out into subcategories along functional lines, such as disease prevention, diagnosis, and screening activities. This is important since, as noted earlier, not all medical spending can be attributed specifically to the treatment of a disease or condition. Whichever method of allocating expenditures is used, it has to offer a solution to the comorbidity problem.

As different output measures are developed, some mechanism will be needed for commissioning demonstration projects to determine what actually makes a difference in practice.

Recommendation 3.3: The Bureau of Economic Analysis (working with academic researchers and with the Bureau of Labor Statistics) should continue to investigate the impact of different expenditure allocation approaches—particularly the episode- and person-based methods—on price index con-

struction and performance. Research is needed to determine which method is best under different circumstances.

As part of this effort, BEA (perhaps in coordination with AHRQ and the National Institute on Aging) should sponsor a workshop for the three vendors of episode grouping software and the top three or four person-based case-mix system vendors to present their products, how they are used in the marketplace, and the underlying rules and logic.

At this point, a cautious approach is warranted, as it is too early for BEA to buy into a particular method for aggregating treatment costs. This means that there may be a need for parallel sets of accounts, at least on a research basis, for some time. BEA researchers should continue to experiment with competing methods (some, such as regression techniques, would be statistical; others may be deterministic) of parsing expenditures into disease groups using different kinds of data (e.g., claims records, survey data). It would be helpful to get a practical sense of how different results would use various approaches and data sources. Results of comparisons will depend on the level of disaggregation. It will be difficult to determine which method, in the abstract, is best—there will inevitably be some joint production with arbitrary allocations of dollars. And there are practical considerations—for example, the proprietary nature of the grouper software—that may steer the work toward particular approaches and away from others.

Recommendation 3.4: The Bureau of Economic Analysis, working with academic researchers (and perhaps other agencies, such as the Centers for Medicare & Medicaid Services and other parts of the Department of Health and Human Services), should collaborate on work to move incrementally toward the goal of creating disease-based expenditure accounts by attempting a "proof of concept" prototype. Using a subgroup of the population with good data coverage, the prototype would attempt to demonstrate that dollars spent in the economy on medical care can be allocated into disease categories in a fashion that yields meaningful information.

Choices will have to be made about how to aggregate rare events and unusual comorbidity combinations. The project should attempt to determine how sensitive expenditure allocation figures are to alternative choices.

Selection of an appropriate group for the pilot should be based on data quality and completeness. The Medicare population, the military, or veterans—groups for which their spending and health data are available (and for whom a good deal of the medical care action takes place)—would be logical choices. Alternatively, a disease-costing pilot could be done for a well-defined, geographically (and administratively) complete group, such as found in parts of Intermountain Healthcare, Geisinger Health System, or one of the Hawaiian islands, before attempting it on a national basis.

4

Measuring Prices and Quantities of Medical Care: Improving Medical Care Price Indexes

A national medical care account requires measures of the quantity of medical care services, as does a national health account. Medical care services are the outputs of the medical care sector, and thus also of the medical care account, and they are one of the inputs into the production of health.

In this report, we have taken the position that the units of medical care production are defined by treatments of disease, or treated episodes of illness, which is also our measure of medical services. Some medical services are not associated directly with the treatment of disease; nevertheless, the disease-based metric provides the framework for measuring most medical services. The many difficulties in estimating expenditures by disease have been discussed in Chapter 3. Here we discuss estimating the quantity of services produced by the medical care sector, as well as the associated measure of medical care inflation.

When estimating the output of medical care, the fundamental problem is measuring the price and the quantity of treatments for a disease—the treatment of heart attacks, for example. Once that is accomplished across the range of diseases, it will be necessary to form aggregate measures of medical services and of medical care inflation. The exposition of this chapter benefits from beginning with a discussion of the second step, forming the aggregations.

4.1. ALTERNATIVE ESTIMATIONS OF QUANTITY GROWTH IN MEDICAL CARE SERVICES: THE ROLE OF PRICE AND QUANTITY INDEXES

Two methodologies exist for aggregating quantity changes of medical treatments. The first, "deflation," borrows the methodology conventionally used by compilers of national accounts: the change in expenditures for some category of goods and services is divided by a price index (a measure of inflation) for that category to obtain the quantity measure.

The alternative method is to aggregate the quantities directly. To do this, a quantity index of treatments is computed (e.g., for heart attacks) using the numbers and types of treatments and their costs, and taking into account differences in severities and in modes of treatment. This approach is less frequently used in the computation of national accounts. However, for countries in which the government provides medical care (so the price charged is not relevant), the direct quantity index approach is an attractive one. Even for the United States, the direct approach has advantages and should be considered, although data for estimating the direct quantity index number approach for medical care are not easy to compile due to the fragmented systems and the difficulty in accounting for variation in severity.

4.1.1. Deflation

Deflation is the standard national accounts approach to obtaining quantity measures. The term "deflation" describes a process in which the central step is dividing the change in expenditures between two periods by a price index. Deflation results in a measure of the growth of quantities.[1] As a medical care example, expenditures on treating heart attacks could be divided by a price index for that service. The result yields a quantity measure—it is an index number—indicating the change in number of treatments for heart attacks. The index also shows the rate of growth in medical services for this medical ailment. Generally accepted procedures for deflation are presented in the system of national accounts (Fisher, 1993, Chapter 16).

The United States and Canada use the Fisher index number system for estimating gross domestic product (GDP). Note, first, that the expenditure on any item equals its price times its quantity, so the expenditure for any group of items equals ΣPQ. Deflation under the Fisher index system takes the following form: change in expenditures / Fisher price index = Fisher quantity index. Algebraically, for periods 0 and 1, the equation is as follows:

$$(\Sigma P_1Q_1/\Sigma P_0Q_0) \,/\, \{[\Sigma P_1Q_0/\Sigma P_0Q_0]\,[\Sigma P_1Q_1/\Sigma P_0Q_1]\}^{1/2} = \{[\Sigma P_0Q_1/\Sigma P_0Q_0]\,[\Sigma P_1Q_1/\Sigma P_1Q_0]\}^{1/2}. \quad (4.1)$$

The Fisher index number is the geometric mean of Paasche and Laspeyres index numbers. Thus, in equation 4.1, the first term in square brackets, $[\Sigma P_1Q_0/\Sigma P_0Q_0]$, is a Laspeyres price index, and the second a Paasche price index. The Paasche and

[1]Often, and misleadingly, it is called by many economists "real GDP," or "real consumption," and so forth. The idea behind the language is that the quantity index measures the growth in actual consumption, or consumption in physical units, as opposed to just the growth in outlays for consumption, which may rise solely because of inflation.

Laspeyres indexes in square brackets are multiplied together and then the square root taken to get the corresponding Fisher indexes, which are represented by the terms within the curly brackets. Thus, the denominator term on the left-hand side of equation 4.1 is equal to the Fisher price index, which equals the square root of the Laspeyres price index times the Paasche price index.

The Fisher price index in equation 4.1 functions as the "deflating price index" or the "deflator." The Fisher quantity index (the right-hand side of equation 4.1) is a parallel construction.

The Bureau of Economic Analysis (BEA) uses the Fisher index number system for GDP as a whole and for sectoral measures. For example, the output of North American Industry Classification System (NAICS) 62 (medical care and social services) in the BEA industry accounts is a Fisher quantity index.

Note that prices are the weights in the Paasche and Laspeyres quantity indexes and hence in the Fisher quantity index. That is, aggregation of the different treatments—for circulatory disease, digestive disease, and so forth—proceeds by valuing the treatments using their prices. Price weighting of the quantity index on the right-hand side of equation 4.1 is the result of the algebra of the deflation method.

The theoretically correct way to aggregate output is by marginal cost (*MC*). By the usual competitive assumption, *MC*s equal prices, which are the weights in the quantity index (the right-hand side of equation 4.1). Thus, deflation preserves (approximately) the theoretical aggregation condition.

The usual national accounts justification for deflation (instead of estimating the quantities directly) is the presumption that prices *within* a category move more nearly together than do the quantities. Thus, the prices of, for example, different kinds of meat or different grades of beef steak are assumed to move together, although the quantities may not. A small sample of prices can therefore yield an accurate price index for the category.[2] A small sample of the quantities, however, is likely to be invalidated by the variance.

The assumption that prices and marginal cost measures are equal is tenuous for any industry.[3] However, it is especially problematic in medical care, for which many prices have remote connections to costs. Even in the United States, therefore, the usual national accounts deflation methodology is problematic when applied to medical care. In other countries, the United Kingdom for example, government-provided medical care means the price index for medical care has no

[2]If all prices within a category move together, Hicks (1940) showed that goods or services within the category could be treated as if they were a single composite good. Sampling for a price index that will be used for deflation thus rests on the Hicksian aggregation theorem. In practice, minimizing price change variance is one principle for structuring samples for price indexes. The statistical methods can therefore be regarded as implementing Hicksian aggregation in the price indexes.

[3]For example, Aizcorbe, Flamm, and Khurshid (2002) evaluate the error in semiconductor output from violation of the $MC = P$ assumption. For semiconductors, the error did not have appreciable impacts but mainly because the costs of semiconductors have fallen so fast that in the time series change in *MC* swamps the error.

relevance. For this reason, unlike in the United States, quantities of medical care in the UK national accounts are not estimated using deflation methods.[4]

One might question as well whether Hicksian aggregation holds for medical care. That prices of alternative heart attack treatments move together over time appears unlikely. For this reason, as well, the deflation approach to estimating medical care output is problematic. Movements in quantities may be equally diverse across alternative treatments for the same disease. However, collecting large samples of quantities of medical treatments is at least as feasible as collecting large samples of their prices and possibly less expensive.

4.1.2. Direct Quantity Indexes

Instead of indirectly computing the right-hand side of equation 4.1 using price indexes (justified by the assumption that $P = MC$), one could calculate the right-hand side of the equation directly, using quantities of treatments and cost estimates. That is (using the Fisher quantity index):

$$\text{quantity index} = \{[\Sigma C_0 Q_1 / \Sigma C_0 Q_0]\,[\Sigma C_1 Q_1 / \Sigma C_1 Q_0]\}^{1/2}. \qquad (4.2)$$

For example, an index would be computed from quantities of treatments; classified by case severity, type of treatment, and so forth; and weighted by the costs (not the prices) of the various treatments. The correct measure is marginal cost; in practice, average cost is more likely to be obtainable and may provide an adequate approximation to marginal cost.

Several European and Oceanic countries are experimenting with direct quantity indexes of medical care. Schreyer (2008) lists Australia, Finland, France, Germany, the Netherlands, New Zealand, Norway, Sweden, and the United Kingdom in this group. In some of these cases, however, the indexes are not computed on a disease basis. Yet price indexes by disease classification (needed for the deflation alternative in section 4.2.1) are relatively rare as well.

It seems advisable to pursue both strategies identified above. We encourage BEA to estimate direct quantity indexes when data can be developed to make their estimation practical. One reason for doing so is that the direct quantity index is the way other countries are likely to proceed, so carrying out similar computations for the United States is important for assessing international comparability. A second reason is that it is by no means clear how long the United States will stand apart from other countries in its method for delivery of medical care. Prices are already not all that relevant to the U.S. medical sector. Costs, however, are

[4] Schreyer (2008) says that the health measure in the UK national accounts is "a cost weighted activity index." However, these UK "activities" are not completed treatments. The UK measure is not an adequate output index.

always relevant and medical care reform is likely to increase the need for efforts to measure them more accurately.[5]

Recommendation 4.1: The Bureau of Economic Analysis should experiment with estimating direct quantity indexes of medical care in addition to its usual deflated measures.

4.1.3. Indexes at Different Levels of Aggregation

We have alluded to the fact that indexes, whether price indexes or quantity indexes, will be computed at different levels of aggregation. An index is also necessary for medical care overall, because the sector involves an aggregation over many different medical conditions. Similarly, an index number is needed to construct a measure for individual types of treatment—for example, circulatory diseases. Such an intermediate-level measure would aggregate treatments for heart attacks, strokes, and so forth.

Even at the lowest feasible level of measurement, indexes are still necessary. For example, many treatments for heart attacks exist; any aggregation of them requires constructing an index number (price or quantity) for this treatment category. An index number is a way of controlling for heterogeneity—differences in treatments, in severity, or in demographics of the patient, as well as other factors. It is hard to conceive of any medical care output that is sufficiently homogeneous that a straight count of the number of treatments, or a simple average of treatment prices (in the price measurement literature known as a unit value index), would be adequate. Either one will be contaminated in the time series by differences in the mix of treatments or severities or other factors that influence both the treatment undertaken and its cost or price (see further discussion of this point in section 4.3).

Put another way, all feasible measurements in medical care are aggregates, because they encounter heterogeneity (in some cases extreme) in the units they encompass. The heterogeneity must be allowed for and controlled in the measurement. Otherwise, a movement toward more effective, but more expensive, treatments may be mistakenly interpreted as an increase in the price of medical care, when it should instead be interpreted as an increase in the quantity of care.

---

[5]A very small literature compares the results of direct quantity indexes with deflated ones. The National Bureau of Economic Research (1980) found quite different trends, no doubt partly because the exercise was carried out on different databases. If Usher's findings extend to medical care, then international differences in national accounts estimates of medical care output growth between the United States and Europe may reflect differences in measurement methodologies. Newhouse (1992) pointed out that growth rate differences in health care expenditures are much smaller than level differences; nevertheless, there is a long and appropriate emphasis on measurement differences in the analysis of international price and expenditure comparisons of all kinds (see for example, OECD study of medical costs across countries).

It was for this reason that this chapter opens with an exposition of index number aggregation schemes. Even though in practice one begins by measuring prices and quantities of treatments for particular diseases, they will still be index numbers. These lower level measurements have similar properties to the indexes that describe GDP or any other economic aggregate.

4.2. PRICE INDEXES FOR MEDICAL CARE: THE PRODUCER PRICE INDEXES AND THEIR USE IN BEA ACCOUNTS

BEA currently uses components of the Bureau of Labor Statistics (BLS) Producer Price Index (PPI), supplemented with some other price measures, to deflate the medical care components of the National Income and Product Accounts. The PPI indexes are used to estimate output for the existing BEA industry accounts for NAICS 62 and its subsectors. As noted in the previous section, when the receipts of a NAICS industry or sector are deflated by a medical care price index, the result is a quantity index of medical services produced by the industry or sector (equation 4.1).

For the PPI, BLS tracks the changes over time in prices received by U.S. domestic producers for their goods and services. As described in Chapter 2, NAICS provides the organizational structure; in the case of medical care, separate indexes are generated for a general hospitals, psychiatric and substance-abuse hospitals, other specialty hospitals, physicians' offices (with component indexes by specialty), diagnostic imaging centers, medical labs, nursing care facilities, residential mental retardation facilities, home health care, and blood and organ banks (all of these are NAICS 3-digit subsectors or 4-digit or 5-digit industries).

For medical care PPIs, the recorded price includes reimbursements to the medical care providers from all sources, including the patient, insurance, Medicare, and Medicaid. Unlike the Consumer Price Index (CPI), government payments and payments by insurance companies are included in the PPI medical care indexes (Catron and Murphy, 1996; Murphy and Topel, 2006). The pricing unit varies by industry. Examples of pricing units are a patient's stay in a hospital for a treatment of a specified diagnosis, from admission to discharge, or the services provided by a physician in one patient visit. For nursing homes, the unit is essentially a day of nursing care.

4.2.1. Hospitals

The PPI hospital methodology, initiated in 1992 with its then-new hospital price indexes, marked a significant advance over what had been done historically. Previously, only the CPI had a medical care component. The PPI covered mostly goods-producing industries. In the old CPI, the unit of measurement was defined by such items as the cost of a day in a hospital or of a visit to the doctor.

The PPI moved to the episodes-of-treatment concept, in which a diagnosis for an in-hospital treatment is priced out for the duration of the inpatient stay. Then, in subsequent periods, BLS asks the hospital what it would charge to treat a diagnosis with the same characteristics—the same severity, the same demographics, and other conditions. Notice that BLS selects a diagnosis in its PPI hospital index sample; it does not sample the treatment, which could change.

When the treatment for the same diagnosis does change, BLS asks the hospital for the cost difference between the new and the old treatments. The cost difference provides the basis for a quality adjustment in the index. For example, suppose that in the initial period the hospital provided a figure of $3,000 as the cost of the existing treatment for a diagnosis. Then, in the subsequent period, a different treatment is used for this diagnosis, and the hospital gives $3,600 as the new cost figure. The hospital, however, also reports that the cost difference between the two treatments (considered at the same time, that is, in a period when both were in use) was $500. In this example, the PPI would record $100—not the full $600—as the price increase for treating the diagnosis.

The PPI index for "general medical and surgical hospitals" is published with category detail that corresponds to chapter and subchapter headings of the International Classification of Diseases (ICD).[6] Of the 20 disease groupings that have been published since 1992, index values in June 2008 ranged from 147.1 for infectious and parasitic diseases (1992 = 100) to 229.6 for diseases of the blood and blood-forming organs and immunology.[7] Hospital costs have not risen at the same rates across diseases. This provides one demonstration why more detail is necessary for analyzing medical care costs and services than is typically provided by the "hospital" aggregate that is published in the National Health Expenditure Accounts.

Several new hospital disease indexes have been added recently, including treatment of trauma and of HIV. Additionally, there are PPI indexes for specialized psychiatric hospitals and substance-abuse hospitals, as well as for residential mental retardation facilities. These indexes map onto the mental health chapter of the ICD.

The improved methodology in the post-1992 PPI hospital index caused it to grow less rapidly than an index using the older method. Catron and Murphy (1996) suggest several reasons for this in addition to improved methodology. For one, their evidence indicated that hospital charges to insurance and other third-party payers advanced more slowly than hospital charges to individual payers, who alone are represented in the CPI (charges and transaction prices can differ wildly and charges certainly need bear no relation to *MC*). Even after the CPI

[6]PPI categories can be found in the table for the net output of selected industries and their products on the BLS PPI web page (see http://www.bls.gov/ppi/ppitable05.pdf). The PPI hospital component also contains an index for "other receipts," which includes revenue from nonmedical operations, such as gift shops.

[7]All these indexes were subsequently rebased to June 2008.

shifted over to a modification of the PPI methodology, PPI hospital indexes have continued to advance more slowly than CPI indexes. Nevertheless, pricing a diagnosis, instead of simply collecting the hospital's daily charge and ancillary charges, was a substantial improvement, and it seems certain that the improvement removed upward bias from the measure of medical care inflation.

As a pragmatic matter, BLS was able to base its PPI hospital indexes on an episode-of-disease concept because hospitals are compensated both by Medicare and many commercial insurers by diagnostic related groups. For this reason, hospitals have the data that the PPI hospital index requires.

4.2.2. Other PPI Medical Care Components

The PPI hospital indexes, including an index for specialty hospitals, are estimated and published using a disease classification. Hospitals account for about 45 percent of the national accounts medical care sector. However in the past, nonhospital PPIs have not been collected on an episode-of-disease concept. Some nonhospital PPIs—for example, physicians' offices of obstetrics and gynecology and offices of mental health practitioners—can be readily mapped into the cost-of-disease framework. Similarly, mental hospitals fit naturally within an ICD chapter.

For other segments of the health care sector, data are not yet available that would work in a cost-of-disease framework. However, BLS recently tested collecting prices from other medical care industries according to the cost-of-disease concept used for the PPI hospital index. These include physicians' offices, medical laboratories, and diagnostic imaging centers. Pharmaceuticals have also been coded by the cost-of-disease classification. These are very positive steps.

BLS plans on grouping these components to achieve ICD indexes that report change in the cost of disease on a wherever-treated basis. That is, in addition to publishing PPIs by industry (e.g., physicians' offices), they will publish indexes by disease (e.g., circulatory disease). At present, they do not plan to publish cost-of-disease indexes for each industry, so there will not be cost-of-disease indexes for, say, physicians' offices, comparable to the ICD-level detail currently published for hospitals.

Recommendation 4.2: If sample sizes permit, the Bureau of Labor Statistics (BLS) should publish not only aggregated (over industries) cost-of-disease indexes but also indexes for each industry (that is, hospitals, physicians' offices, and so forth) by cost-of-disease classification, so that users can examine the components of disease cost and do their own analysis. The panel is encouraged by BLS plans to extend the cost-of-disease framework used in its hospital Producer Price Index to other medical care indexes, including doctors' offices and other ambulatory care industries. This initiative should be carried forward and given appropriate funding.

4.2.3. Deflation with PPI Indexes

Even though PPI hospital indexes have been available with a cost-of-disease classification for more than 15 years, BEA does not deflate at the disease level. No expenditure data are available by cost of disease, so there is nothing for BEA to deflate. For that reason, BEA uses the aggregate PPI for hospitals as the deflator. This situation underscores how inadequate are data for the medical care sector, compared with other sectors of the economy. In other sectors, PPI indexes usually provide less detail than is available for expenditures, so it is typically the availability of PPI indexes that limits deflation detail, not the availability of expenditure data.

As noted in Chapter 3, the 2007 Economic Census collected revenue from hospitals and from certain other medical care industries by the same diagnostic-related groups classification system used in the PPI. This welcome development was the result of the North American Product Classification System (NAPCS), which developed harmonized product classification systems for use in Canada, Mexico, and the United States. For the United States, the NAPCS was also intended to harmonize classifications used by the Census Bureau and BLS in order to facilitate analysis by users of government statistics and deflation by BEA. The NAPCS specified product detail in medical care that is consistent with the ICD classification.

It is obvious from equation 4.1 that the units of analysis must be the same for the price and expenditure data used in the equation. Lack of product classification consistency across U.S. statistical agencies has in the past forced various expedients upon the agency responsible for deflation. In the immediate future this problem will be resolved by new data from the Census Bureau and from the PPI. Provided that yearly extrapolators of the 2007 census product detail are available, the U.S. statistical system will have gone a long way toward providing expenditures and price indexes for medical care that are harmonized around a cost-of-disease framework.

Recommendation 4.3: The Census Bureau should give high priority to providing annual data for hospitals and other medical care industries, grouped by a cost-of-disease system that matches the one used in the 2007 Economic Census and in the Producer Price Index. The panel notes the welcome provision of new data on receipts categorized by disease that accompany publication of the 2007 Economic Census, though a considerable amount of work remains to be done to bring the quality of these data up to the standards needed for use in official statistics.

Specifically, a concerted effort will be needed to increase the number of census respondents willing to provide information at the disease level of detail.

4.3. WHAT IS NEEDED IN A PRICE INDEX FOR MEDICAL CARE?: EVALUATION OF THE PPI

The emergence of price indexes and expenditure (receipts) data that are grouped by a cost-of-disease system makes it essential to examine the existing price indexes, to assess their adequacy for measuring output and inflation in the medical care sector.

BLS has announced its intention to extend the PPI hospital methodology to other indexes. Thus, rather than assessing the existing BLS nonhospital indexes, we examine issues that have arisen in the PPI hospital index. The discussion will apply by extension to indexes for other areas of care or to their development.

4.3.1. Synthetic or "Model" Prices

For most parts of the PPI, BLS uses a fixed-specification pricing methodology in which price comparisons are made only for units that are matched in two periods. For the product or service that was selected in the initiation period (generally by probability sampling), BLS obtains a price in subsequent periods for the exact specification, in order to hold product or service quality constant in comparing prices. For medical care, however, it is very unlikely that a condition selected (on a probability basis) in an earlier period will present itself subsequently in exactly the same form. The sampled price for heart attack treatment one month may be for a 55-year-old male with particular form of heart attack; the next month it may be an 80-year-old with another form. Thus, strict application of the usual fixed-specification pricing method will not work well for pricing medical care.

In order to get repeat price quotes for a fixed diagnosis, BLS obtains a synthetic price for the hospital index. That is, it asks the hospital what it would have charged in the current period for the exact diagnosis that BLS selected in the initiation period, controlling for severity, demographic profile, and other relevant factors (Murphy workshop presentation, in National Research Council, 2009). The method currently used does not adjust for quality by tracking outcomes (although the framework itself would be amenable to that), and it could allow for changing technologies. The synthetic price approach employed for the hospital index (which will presumably be used for the doctor's office and other medical care indexes planned for the future) was adapted from the "model price" method developed by Statistics Canada for construction price indexes.[8] BLS has also applied the model price method to various services industries such as engineering, for which finding the exact item to price in subsequent periods is impossible.

Note that BLS could have estimated a unit value index for, say, heart attacks, but it did not. That is, it could have taken a simple average of the charges for

[8]However, Statistics Canada has subsequently abandoned this method for pricing construction projects.

all heart attack treatments in the hospital in the initiation period, another simple average of the charges for all heart attack treatments in the subsequent period, and let the ratio of those two simple averages form the price index. Such a ratio is called a unit value index in the price index literature. A unit value index implies that all treatments are homogeneous or that any heterogeneity among them can be neglected. Changes in the mix of treatments or in the characteristics of the patient either do not matter or are small enough to be ignored.

On the quantity side, the unit value index implies that a simple count of the number of, for example, heart attack treatments suffices. Bearing in mind the definition of the unit value, $(\Sigma PQ / n)$, where n is the number of treatments in the period,

$$\text{unit value index} = (\Sigma P_1 Q_1 / n_1) / (\Sigma P_0 Q_0 / n_0), \quad (4.3)$$

where n_0 and n_1 are the numbers of treatments in periods 0 and 1, respectively. Deflation by the unit value index gives the following:[9]

$$\text{output change} = \{(\Sigma P_1 Q_1 / \Sigma P_0 Q_0)\} / \{(\Sigma P_1 Q_1 / n_1) / (\Sigma P_0 Q_0 / n_0)\} = n_1 / n_0. \quad (4.4)$$

Thus, if the unit value index is valid, there is no need to control for the severity of cases or for any other characteristic that might affect treatment. If the unit value index is valid, computing the quantity index is very simple because it is necessary only to count the number of treatments, without controlling for patient mix, severity, or any other characteristic of the patient. And, there is no need to collect the costs, either for deflation or for weighting the quantity index: if the number of treatments is available in both periods (needed in any case to compute the unit value index), there is no need to deflate by the unit value index; one can just use the change in the number of treatments.[10]

Extreme heterogeneity in medical treatments, however, exists. From a treatment perspective, rarely do two patients have exactly the same illness. And the distribution of patient severity may change over time. For example, as some surgical cases have shifted to ambulatory surgery and out of the hospital, the remaining inpatient cases were more severely ill. Heterogeneity makes problematic both the normal BLS fixed-specification method used for other goods and services as well as the unit value index method. By default, therefore, BLS adopted the approach of estimating a synthetic price for the exact specification that was chosen in the initiation period.

[9]See also Balk (1998), who presents the same result.

[10]The statement needs qualifying because the price measure and the expenditure measure are frequently computed from different databases (for example, from the Annual Surveys of Services and the PPI).

The PPI synthetic price method worked, in the sense that it proved to be practical and gave plausible indexes. However, no real evaluation of the method has ever been carried out, and it is warranted. How do hospitals answer the BLS hypothetical question? Do they just quote from some standard charge list? Do they mark up costs in some way? Do they take shortcuts, such as looking at changes in compensation of personnel? In short, how valid are hospitals' responses? It seems impossible to make any kind of judgment about the validity of the PPI medical price indexes without knowing more about how the respondents determine their answers to the BLS question.

Recommendation 4.4: The Bureau of Labor Statistics (BLS) should undertake a special study of how hospital respondents estimate their current charge for the detailed specification BLS chooses in the initiation period.

BLS survey statisticians have in the past performed "response analysis surveys," which examine how respondents reported information to BLS. An example of such a study is Goldenberg, Butani, and Phipps (1993). The proposed study should also encompass the way hospitals have perceived and responded to BLS instructions about changing treatments, to determine why few treatment changes have been reported, along the lines suggested in section 4.3.3.

4.3.2. Quality Change Adjustments for Improved Treatments

It is generally accepted, both in the price index literature and in medical economics,[11] that price indexes need to be adjusted to reflect improved medical treatments or (as it is usually called in the price index literature) quality change. Fisher and Shell (1972) and Triplett (1983) show that the appropriate way to make a quality adjustment in an index of output (or in the output price index) when markets are competitive is by the cost difference between the older and the improved treatments. BLS uses the change in cost as a quality adjustment when changed treatments are encountered in the PPI hospital index. An example has already been given in section 4.2.

Although the BLS quality adjustment method corresponds to the dictates of the theory, we note that the conditions of the theory are very restrictive. The restrictions make it problematic when applied to medical care. The theory of the output price index, adopted as the theoretical framework of the PPI, is the theory of a constant input, fixed technology price index.[12] It is isomorphic to the theory of the cost-of-living index, which is a constant utility, fixed preference function index. In that fixed input, fixed technology world, the theoretical ideal is to use the difference in production costs between the old treatment and

[11]See the instructive analysis from the medical economics side by Phelps (1999, pp. 108-117).

[12]See Fisher and Shell (1972) and also Archibald (1977).

the new treatment to make a quality adjustment in the index (because otherwise the index would not hold inputs fixed).

An input-cost adjustment is warranted if the quality change does not involve a shift in the underlying production technology. Some quality changes fit this model. For example, computers have over some periods used the same underlying technologies, but improvements make machines faster. Greater hospital resources put into cleaning and sanitation in order to reduce in-hospital infections fits the fixed technology model as do some of the quality improvements in treatments distinguished in the Hospital Compare Project (administering aspirin to heart attack patients, for example).

In medical care, however, many treatment changes involve new technologies. The cataract surgery that moved to a sutureless procedure is a good example (see the quality change section below). It was not a constant technology innovation. This innovation not only improved the treatment but actually reduced its cost, so basing the quality adjustment on the cost change is inappropriate. In addition, it probably improved treatment outcomes further by reducing the likelihood of mistakes and complications from putting in and removing the suture. Treating complications adds to cost, which should be picked up in the price index used for the accounts, although to do so would require that the cost of complications from a procedure be linked to the original episode of treatment. When these costs are eliminated, it should count as a productivity improvement, which would have to enter the accounts through an adjustment to the price index.

In principle one could ask, what would it cost to produce the characteristics or outcome of the new treatment in the old technology? However, the outcome of the new procedure often cannot be produced using the old technology. One could also in principle ask what it would cost to produce the characteristics of the old treatment with the new technology, but that is a nonsense question in the cataract surgery case, since the new sutureless cataract surgery technology so completely dominates the old.

Cases such as those described above demonstrate that the only clear alternative is to revert to medical outcome measures, even though outcomes do not, strictly speaking, comport with the theory that underlies the PPI. This is not an easy solution, either, since outcome measures are few. The role of medical outcomes in national health accounting is twofold: (1) it is required for tracking quality change in the output (the treatments) for the medical care account,[13] and (2) it is (as we discuss in Chapter 6) central to the broad health account, since changes in population health need to be attributed categorically to specific treatments of disease.

[13]A detailed discussion of BEA and BLS plans for improving medical care price indexes and for dealing with quality change can be found in National Research Council (2009).

4.3.3. Frequency of New Treatment Encounters in the PPI

BLS has said that few hospitals report treatment changes for the hospital indexes (Fixler and Ginsburg, 2001).[14] This is somewhat puzzling, given the frequent and often major changes that have taken place in medical practices over the years. One might expect that PPI collection would encounter treatment changes frequently, not infrequently. One possibility is that the BLS pricing mechanism discourages reports of changed treatments. Because BLS collects a synthetic price, as described in section 4.3.1., the hospital need not report a treatment change to continue to report to BLS. In principle, that is, BLS tracks a diagnosis, not a treatment, but the hospital may find it easier to report a fixed treatment, despite BLS instructions. However, little is known about this beyond the useful discussion in Fixler and Ginsburg (2001). Clarifying what the hospital is reporting and how it interprets and responds to BLS instructions is part of the research in Recommendation 4.2; the study should shed light on whether the low rate of treatment changes reported by hospitals to BLS is erroneous.

A second possibility combines to an extent with the first. BLS selects for its hospital index a sample of diagnoses from a hospital, and retains the sample for 7 years. Perhaps the BLS repricing cycle is too short to detect many treatment changes or (more likely) too short for an old treatment to disappear fully, even if a new one has been adopted. If the old treatment is not completely supplanted by a new one, the hospital may find it easier to report data on the old one (the respondent likely knows that if a treatment change is reported, that person will be asked for more information, which means more work).

Third, the encounter rate for new treatments that BLS should expect is unknown. It is known that, over a long time period, treatments for nearly every medical condition change, as does the mix of characteristics and ailments of the patients being treated (Fuchs, 1999). But what *proportion* of treatments might be expected to change in a relatively short interval? If diffusion studies of new treatments, comparable to the famous study of hybrid corn (Griliches, 1957), were available, they would be a better basis for judging whether the rate at which BLS encounters new treatments is really low.[15]

Recommendation 4.5: The Bureau of Labor Statistics (perhaps with the help of outside researchers) should evaluate the implications of the apparently low rate of encounters with new treatments in its hospital indexes. Examining data from reimbursement protocols could provide a benchmark for how rapidly treatments are changing in medical practices. It is likely that a detailed sample of some medical care components is an appropriate approach. Select-

[14]The information given by Fixler and Ginsburg applies generally to the PPI (conversation with BLS staff, March 2009).

[15]As an extra complication, what is a "new" treatment? How much change in process or protocol leads the hospital that reports to BLS to decide that a new treatment has been introduced?

ing those International Classification of Diseases chapters or subchapters (such as the circulatory diseases chapter and the nervous system and sense organs chapter) for which existing research has shown rapid technological changes in some treatments has the advantage of placing the benchmark in the context of the economic research that is most often cited.

4.4. WHEN TREATMENTS MOVE ACROSS INDUSTRIES OR ACROSS ESTABLISHMENTS

We note in Chapter 2 that completed treatments may extend across industry lines—hospitals, doctors' offices, clinics, laboratories, and skilled nursing facilities are all in different NAICS industries. That mix of facilities poses problems in obtaining the cost of a completed treatment, although we concluded that the problems were relatively manageable.

However, another type of shift between and among different medical establishments is more problematic. Shapiro, Shapiro, and Wilcox (2001) explain that cataract surgeries were once performed in hospitals but are now an outpatient treatment, performed at much lower cost. Moreover, in terms of outcome, the lower cost treatment gives as good an improvement in vision as the more expensive hospital treatment of earlier days, and the authors contend that the less expensive treatment is undoubtedly better for having fewer side effects, particularly discomfort in recovery. Changes in the treatment of cataracts provide an archetype for a broad class of changes in medical treatments that pose severe problems for measuring medical care output and inflation.

Under current PPI procedures for pricing cataract treatments, BLS might sample surgeries that take place in a hospital; the resulting price index for cataract surgery would feed into the published hospital PPI for ICD Chapter 6, Diseases of the Nervous System and Sense Organs (code 366 is cataracts). If surgeries shift to a clinic, then another cataract surgery price index would be obtained, which would be, in principle, a component of the ambulatory care index for ICD Chapter 6.[16] Suppose neither hospital nor clinic changed its price. If patients switch from the more expensive hospital surgery to a less expensive ambulatory care facility or a doctor's office, and if quality remains constant (i.e., if treatment outcomes are comparable), then the cost saving to the patient would be missed in the PPI, even if, as BLS is now planning, the hospital and ambulatory cataract surgery operations are aggregated into an overall index for ICD Chapter 6.

The PPI records prices received by each provider, and, in the example, no provider's price changed. But if we are interested in the price paid by the patient, by the insurance company, or by Medicare, the price has fallen. The PPI would not pick up this price reduction.

[16]As noted above, BLS is not currently planning on publishing at this level of detail, but it is implicit in the methodology.

The PPI is constructed on its own concept, an industry price index, and, in the example, none of the industry price indexes is missing anything. But as Berndt et al. (2000) point out, the United States has no national price index for medical care. When a cheaper treatment becomes available in another industry and patients (or their insurance companies) cross industry lines, an aggregation of PPI indexes will not produce an accurate national measure for medical care. What is needed is an index that is capable of capturing the facility shift as a decrease in price.

The cataract surgery problem has been described as a substitution, and substitutions are well known in the price index literature. Another example is that of ulcer treatment moving from a surgery solution to an antibiotics treatment. In either case, the key to understanding the problem is recognition that this substitution is toward the same product at a lower cost; it is not the usual price index "substitution bias" created by shifting demands toward products that have lower relative prices over time.

When medical treatments cross industry lines (the prototypical cataract surgery problem) the problem posed for medical care price indexes is identical, conceptually, to the "discount store problem" much discussed in the CPI literature: pricing products within a store, as BLS does, misses price changes that a consumer experiences from shopping at different outlets (Reinsdorf, 1993; Reinsdorf and Moulton, 1997; Feenstra and Shapiro, 2003). In this case, BLS must compare prices across retail outlets in a way such that the portion of the cost reduction (if any) that is a true price change and the portion (if any) associated with a change in retail services can be estimated. Similarly, for BLS to compare medical treatments across providers in the PPI requires a way to estimate whether the medical outcome is the same when the shift across providers occurs, to partition any reduction in cost into reduction in price, if any, and reduction in outcome, if any. It is clear that some reductions (and possibly some increases) in medical care costs stem from changes in medical care services and are not price decreases.

Some context is needed. In medical care examples, such as the cataract surgery case, the shift in treatment is often a shift away from establishments in one industry (hospitals, NAICS 622) toward others in the ambulatory care subsector of the NAICS (for example, NAICS 621493, freestanding ambulatory surgical and emergency centers). But it need not be. The PPI would also miss price change if one hospital began to offer cataract surgery on an outpatient basis and patients shifted away from hospitals that offered only the conventional treatment with an in-hospital stay.[17] The phenomenon requires a shift in treatments across establishments toward lower cost establishments. They need not be in different NAICS industries, although in the medical cases they often are.

[17]If the shift toward the cheaper outpatient treatment occurred within a reporting hospital, BLS methods should record it as a price decline, provided the hospital reported the treatment change (Murphy and Topel, 2006), but see section 4.4.3.

The cataract surgery problem has been much discussed recently, but usually in anecdotal terms. Missing so far are quantitative measures. How much of the changes in medical care is characterized by treatments moving toward cheaper yet equivalent treatments? For example, it is unknown even how extensive are these types of substitution changes in ICD Chapter 9 (code 366). And how fast are the changes proceeding, and over what interval? A shift that proceeds within a short time period has much different measurement implications than one that proceeds over many years. How many of these shifts occur in institutional settings where conventional collections miss them? Shifts toward lower cost treatments within hospitals, for example, are picked up in principle by BLS sampling methods for the PPI. Are there also treatment shifts that are cost saving but—unlike the cataract surgery archetype—have worse outcomes?

4.4.1. Approaches to the Cross-Industry Problem

Two empirical approaches seem available. The researcher might gather data from individuals or from claims, computing the total cost of a treatment in two periods. Alternatively, an adjustment might be applied to PPI price quotes across industries or in aggregating PPI indexes across industries. The two alternatives have offsetting empirical advantages and disadvantages.

Suppose a new outpatient treatment becomes available and is potentially a replacement for a hospital inpatient treatment. Minimally invasive surgery is an example. If BLS staff knew about the change (see the discussion in section 4.3.3 of the low rate of treatment improvements reported to the PPI), they could in principle collect the price of the new outpatient treatment and match it to the former in-hospital treatment. Thus, using i and j to designate the patients in the two periods and k and m to designate the establishments doing the surgery, and using 1 and 2 to designate the periods, the price relative[18] would be: $p_2 j_2 m_2 / p_1 i_1 k_1$. The notation emphasizes that not only do the periods and the establishments differ, but also the patient and the patient's characteristics do.

BLS would need, in addition, information about the outcomes of the two treatments, because they would need to make a quality adjustment for any direct comparison across different establishments. Hence, the correct price comparison is

$$\text{price change} = \{(p_2 i_2 j_2 / p_1 i_1 j_1) / (\mu_2 / \mu_1)\}, \tag{4.5}$$

where μ is the outcome measure (valued in dollar units and adjusted for any changes in the severity of the patients being treated), subscripted for 1 (old treatment) and 2 (new treatment). Most of the time, the values of μ are unknown or

[18]A price index is computed from matched observations priced in two periods. The ratio of these two prices—that is, the price for the same commodity in two periods—is called a "price relative" in the price index literature. Note that this is not the same thing as the economics term "relative price," which describes the ratio of prices for two commodities in the same period.

must be assumed. If it were known that the minimally invasive surgery was at least as good as the in-hospital surgery, for example, then a direct price comparison without an outcome measure would provide an upper bound on the index, that is, it would understate the price decline. The cataract surgery study by Shapiro, Shapiro, and Wilcox (2001) used this assumption.

It has sometimes been stated that omission of the medical outcome measure may reduce the accuracy of the index, but at least the price change is in the proper direction. But it is clear from manipulation of equation 4.5 that whether the index moves in the right direction depends on whether:

$$(p_2 i_2 j_2 / p_1 i_1 j_1) > (\mu_2 / \mu_1) \text{ or} \quad (4.6.a)$$
$$(p_2 i_2 j_2 / p_1 i_1 j_1) < (\mu_2 / \mu_1). \quad (4.6.b)$$

One can count on the "proper direction" presumption only if condition 4.6.b obtains.

Suppose, however, that BLS ignores the introduction of the noninvasive techniques, that it does not compare prices across industry lines (the current situation) because it lacks measures of μ. Suppose, additionally, that a researcher found or estimated industry values for $p_2 i_2 j_2$, $p_1 i_1 j_1$, μ_2, and μ_1 ; that is, the researcher has estimated how much cheaper was the invasive surgery than the in-hospital kind and how much better or worse was its outcome. In principle, this information could be used to correct for the BLS omission when the indexes for industries j_1 and j_2 are aggregated into the price index for medical care.[19]

This implies a major research agenda. Figuring out how to do the quality adjustment requires much scientific and medical information. Moreover, in many cases an improvement in medical outcomes will need to be traced to multiple sources that apply to different industries. For example, people with bad hips are better off today than 25 years ago, probably mostly because of innovation in the inpatient setting (including devices), but also probably because of improvement in the surgeon's techniques (which does not show up as a hospital charge), better anesthetic technique, and better rehabilitation techniques after discharge.

In a workshop presentation by Bonnie Murphy of BLS (see National Research Council, 2009), it was noted that the PPI program can, in principle, handle treatment shifts within providers. So, if a cataract surgery was changed from a sutured to a sutureless procedure, and it was performed by the same kind of provider—say, in the hospital, even on an outpatient bases—that shift could be captured. If the nonsuture cataract surgery was performed in a physician's office, that would not be, as BLS current index sampling procedures cannot accommodate changed treatments that cross providers. Murphy stated that BLS would in principle want to be able to measure price change associated with these kinds of treatment changes.

---

[19]The nature of the adjustment will depend on the case. An example of using research findings to adjust a medical care price index appears in Triplett (2001). As noted above, the industry price indexes are correct, as industry output measures; it is only their aggregation that poses a problem.

4.4.2. Using Claims or Household Data

Much of the academic literature has relied on patient claims data to provide a picture of price trends for treating specific conditions. In these studies, the good (or service) has been defined, as in this report, as a completed episode. For example, for a heart attack patient, this may involve time and expenditures on a series of initial treatments plus those that take place during the recovery period At the end of that episode, an estimate of all dollars spent for a patient over the entire period is collected; this forms the basis for pricing a completed episode.

Claims data minimize many of the difficulties with provider-side data. Provided patient links are retained across providers, claims data record the cost of the whole patient episode, and there is no need to join together data from different providers—that is, in the notation used previously, one starts with the observation: $\sum_j c_{ij}$.

Moreover, claims data are available in very large sample sizes. BEA has been working with a claims data set from insurance companies with 700 million observations. The heart attack study by Cutler et al. (2001) used over 1.7 million Medicare records (they had additional records from a major teaching hospital). In contrast to those magnitudes, the PPI samples appear minuscule: Fixler and Ginsberg reported that the PPI hospital index sample in 1997 consisted of 1,602 price quotes from 209 establishments (down through attrition from 358 at initiation). At the same date, there were 761 quotes in the PPI physicians' index (Fixler and Ginsberg, 2001, pp. 231, 240). BLS drew new samples in 2001, but no information is available on sample sizes. Clearly, the PPI samples are not nearly so comprehensive as those drawn from claims data.[20]

Offsetting their undeniable advantages, claims data usually require computing unit values as price indexes. For some databases, there is no basis for matching observations in adjacent periods; however, some longitudinal claims databases do exist, and more may be created. Thus, researchers may increasingly have opportunities to compute the ratio of average prices per patient for two periods.

4.5. UNIT VALUES COMPARED WITH SPECIFICATION PRICING

A unit value is simply the value for some category divided by a count of the number of units in that category. For example, unit values for imported autos are derived from the total value of autos imported in some time period divided by the number of cars imported; the unit value index is the change in unit values from one period to the next.

[20]We are aware that sample size, especially a very large sample, is not everything. In particular, the PPI sample is drawn on the basis of a very sophisticated probability design, in which not only selection of responding units but also the procedures chosen for pricing are selected to ensure national representativeness. The claims data that have been employed in most research studies do not have the assurance of representativeness. Nevertheless, for some of the analytic purposes that are vital for medical care studies, large samples permit analysis of detail that would not be feasible with samples as small as those used in the PPI.

A specification price index for imported autos (for example, the BLS import price index) is based on matching cars in samples for two periods, so that the price change is measured only for a fixed specification. A change in the mix of cars imported will affect the unit value index (more economy cars would lower the unit value index, fewer of them would raise it) but would not affect the fixed-specification index. Of course, the fixed-specification index is not without its own problems: if the car in the sample changes its specifications (quality change), some adjustment must be applied in order to factor out price change from value change because of changes in quality. But few international economists would favor the unit value index over the fixed-specification price index, even if they point to measurement problems with the latter (a recent paper on measurement problems in BLS import and export price indexes is Feenstra, Diewert, and U.S. Office of Prices and Living Conditions, 2001).

For collections from patients, it is almost certainly the case that price indexes will be unit value based. Repeat collections of the price for treating a heart attack, with specified conditions including case severity, demographics, and so forth, are the norm for the PPI index for hospitals. Repeat collections from heart attack patients are not a feasible collection strategy since the same patient typically does not go in for a heart attack treatment each period that a price index is constructed (rather, it will be a different person with a heart attack in the next period), so unit value indexes are the only options. Like other unit value indexes, those created for medical conditions will change with the mix of patient characteristics.

We noted earlier (section 4.3.1) the disadvantages of unit value price indexes—that they treat all observations as equivalent and thus variation among observations is ignored. The implied quantity measure is simply the (equally weighted) change in the number of treatments.

When samples are very large, it may be possible to stratify by some variables, such as severity, for example, to attenuate this problem—although claims data generally lack the clinical detail that is often necessary for an adequate adjustment. One need not compute a unit value index for all heart attacks; if data were available, classes of heart attacks could reduce heterogeneity to an extent. Indeed, a BEA study (Aizcorbe and Nestoriak, 2008) distinguished nine heart attack indexes, depending on the type of treatment received (e.g., acute myocardial infarction with coronary artery bypass graft, acute myocardial infarction with angioplasty). Nevertheless, a unit value index implies that the quantity change within the stratum is simply the change in (the count of) the number of treatments.

How serious is unit value bias?[21] What can be said about the empirical liability of the unit value index, compared with the empirical limitations of the

[21] Balk (1998) defines unit value bias as the difference between the unit value index and a superlative index number. However, he is mainly concerned with the form of the index number and does not explicitly discuss undetected mix and quality variation within the unit value index; that is, he writes of the situation in which the units are homogeneous and not of bias created from nonhomogeneity of the units included in the unit values.

provider-side alternatives? In thinking about these questions, recourse to other situations in which unit value indexes are used is valuable. Houses, like medical treatments, cannot usually be priced with the repeat-sale, matched specification methodology employed in price indexes. The change in the average sale price for houses is frequently published. That is a unit value index. It provides valuable and suggestive information and is often based on large samples, but it is well known that the average sale price varies with changes in the mix of houses sold, even when stratified tightly by geographic area. A hedonic price index for houses is an alternative to the unit value index. The hedonic index controls for changes in the mix of houses sold by holding constant the characteristics of the houses, thus providing a more accurate price index. Hedonic indexes for houses are published in many contexts in economics and property appraisers' professional literatures.

Pharmaceutical research provides a medical example. Patricia Danzon, in a presentation to the March 2008 workshop (National Research Council, 2009), noted that prices for pharmaceuticals have often been estimated simply by dividing expenditure by number of units. This amounts to a unit value index. In cross-national comparisons, such unit value data have led to the inference that pharmaceutical prices are much higher in the U.S. health care system.

Pharmaceuticals are precisely defined—they are measured at the level of the mechanism of action, the strength, the pack, the manufacturer, etc. This allows researchers to calculate accurate geographic comparisons of utilization and price differences. Danzon and Furukawa (2006) found that a significant portion of the expenditure difference across countries is explained by variation in the drugs being used—the formulations have quality dimensions to them. Simply dividing expenditures by number of prescriptions can vastly overstate price differences. Using the number of prescriptions in the denominator essentially imputes all the expenditure difference across countries to a price change, whereas much of it is in fact attributable to new drugs or new formulations and to generics. Even with very large samples, international comparisons of prices using unit value indexes will give misleading results. Thus, pharmaceuticals are a good example of why accounting for quality differences is important in medical care price indexes.

There is justifiable concern about the price index bias that arises from missing treatments for disease that cross industry lines (e.g., the cataract surgery example). At this point, not enough is known about the prevalence of such cases, especially in the short run, to do more than speculate about the magnitude of this bias. Even so, moving to alternative data sets that minimize the cross-industry bias is attractive.[22] The more severe the potential bias is from this source, the more likely that one would accept a method that suffers from unit value bias.

[22]Recall the distinction drawn in Chapter 2 between treatments that may bridge several industries and may shift among them and the cataract surgery type of problem in which the same treatment, or one with equivalent outcomes, shifts from a more expensive to a less expensive industry provider. The distinction is equivalent to the distinction between the normal CPI substitution bias (which is resolved with superlative indexes and current weighting) and the CPI discount store bias.

5

Defining and Measuring Population Health

5.1. INTRODUCTION

5.1.1. Motivation

The novel challenge of a national health account is measuring health. In order to answer the question "What are people getting for their health care dollar?" it is necessary to be able to track the health of the population and its subgroups accurately, including those in vulnerable segments of the population. The broad health account requires data on medical care expenditures (and other nonmedical and nonmarket health-affecting inputs) and on the health benefits derived, which are what patients and, collectively, society seek to purchase. The output side of the account is quantified in terms represented by the population's health. Monitoring changing population health on a disease-by-disease basis is also relevant to medical care accounting in the National Income and Product Accounts (NIPAs) because, as discussed in Chapters 2 and 4, tracking health outcomes will ultimately play a key role in quality adjusting the price of medical treatments. Data on health inputs and outputs are also crucial for researchers attempting to link the two sides of the equation—that is, to attribute deaths and impairment to diseases, medical conditions, and other causal factors.

In a satellite health account, output associated with investments in health should be measured independently of inputs. Previous chapters of this report have discussed options for measuring the output of medical care (the treatments), which is an input to health. Independent measurement of health, however, means going beyond simply adding up the value of inputs to yield a value for the output side of the account. In estimating that value, both components of output—the consumption flow of good health and the additional (or reduced) income that a healthier (or less healthy) population generates—should be measured (National Research Council, 2005, pp. 131-132).

In this chapter, we describe and assess various approaches for measuring population health (mortality and nonfatal/impairment dimensions), acknowledging that best measures may vary by purpose and for different populations along the health spectrum. We focus on generic health here, realizing that there is also interest in disease-specific indicators. Our recommendations in Chapter 6 for data collection on major chronic diseases to facilitate research on spending and health linkages would provide the data for a more detailed annual report, as recommended in the Institute of Medicine's *State of the USA Health Indicators* letter report (2009).

In addition to current health status, we also consider risk factors that impact future health. Current health status does not capture the health effects of such risks as hypertension, which does not cause symptoms today but may do so in the future. In principle, these risk factors, together with age, current health, and a variety of other determinants (see Chapter 6) go into health capital (a stock of health), which can be defined as the present discounted value of future health consumption flows (Grossman, 1972). Some portion of a person's health capital is determined genetically, but health can also be affected through investments in inputs ranging from medical care to personal behaviors, such as consumption habits and exercise. While calculating national health capital is not feasible at this time, data on some of the risk factors that go into those calculations could be collected—or, at least, data that are already collected could be extended and more systematically compiled period by period.

Monitoring the population's health status requires metrics that combine quantity and quality of life such as quality-adjusted life years (QALYs). A number of such measures are already widely used to identify unmet health needs and to guide policies for addressing those needs. The multiplicity of measures reflects the lack of an undisputed definition of and method for measuring population health (Kindig and Stoddart, 2003; McDowell, Spasoff, and Kristjansson, 2004).

Nonetheless, some broad generalizations are possible. For the purpose of developing a health account used to evaluate productivity of direct medical and public health services, the health of a population or subgroup within the population is taken to be some combination of survival probabilities and the sum (or, equivalently, the average) of the health of survivors in the population or subgroup. Measuring the health of the population under this assumption reduces to calculating death rates, measuring the health of individual survivors, and aggregating across individuals.

5.1.2. Mortality and Life Expectancy

The oldest and simplest measures of population health are death records. These were used, for example, to calculate the impact in excess deaths of the great plague of 1665-1666 in London (Champion, 1993). With the decline in deaths from infectious disease in the developed world, crude death rates are primarily determined by the age composition of a population. For example, about 0.4 percent of the Mexican population died in 2003 compared with 1 percent of the Italian population. The major reason for the difference was age, with 6 percent of Mexicans compared with 19 percent of Italians age 65 or older (mortality rates at any given age were lower in Italy than in Mexico).[1] So, despite mortality's importance, measures of it must be standardized to reflect age composition in order to be useful for comparative purposes. One possibility is to pick a baseline, such as the year 2000 U.S. population, and calculate directly standardized death rates for all other populations of interest (subgroups, other countries, other years, etc.). These rates are the deaths that would occur if the people in the standard population died at the same rate as those of each age in the population of interest. However, life-expectancy methods provide an approach to standardizing death rates that avoids picking a particular reference population; because of this, it is the most widely used method of summarizing population mortality.

Typically, life expectancy is computed using a period life table, which summarizes the age-specific mortality experience of a population over a short time, usually 1 or 3 years. The National Center for Health Statistics (NCHS) receives mortality information for all deaths in the United States through a cooperative reporting program involving all states and territories. Age-specific mortality rates are computed by dividing the number of deaths of persons at a specific age by the estimated midyear age-specific population provided by the U.S. Census Bureau. From these mortality rates, life expectancies can be simply calculated.

Despite their name, life expectancies are not primarily estimates of the future experience of individuals of a given age. Instead, they provide a method for summarizing mortality experience in the population during the period for which the deaths used in the computations were observed. They are the average number of years a hypothetical cohort of people with a particular starting age who have the current age-specific mortality rates at each future age would live. By contrast, a cohort life table gives the mortality experience of a fixed cohort of people—for example, deaths in a given year, until the entire cohort has died. So, for a cohort born in 1950, the death rates at ages 30 and 50 would be based on death rates observed around 1980 and 2000.

Only under the heroic assumption that age-specific mortality rates will not change for the next 100 years do life tables predict how many years that newborns will live. Because age-specific mortality rates have fallen steadily and are expected to continue to fall, current age-specific life expectancy should underestimate the expected years of survival for an individual now that age. Analysts who need to predict future survival accurately, such as Social Security actuaries,

[1] See http://earthtrends.wri.org/pdf_library/data_tables/pop2_2003.pdf.

must focus carefully on assumptions about how fast age-specific mortality will fall in future decades to obtain good predictions.

Because life tables use only mortality rates, they are not affected by the age composition of the population. For many purposes this is desirable, but some statistics, such as per capita medical spending and death or disability rates, depend heavily on a population's age composition. And for some planning and budgeting purposes, it is necessary to collect or predict actual mortality, disability, and spending, rather than age-adjusted statistics. Analysts must be consistent in standardization for age when comparing trends or levels in expenditures and health. Stratification is a good alternative to standardization and would also prevent misleading confounding-by-age differences in comparisons. One could include either total or per capita health and expenditure flows stratified by age to adjust for population heterogeneity for a given disease.

Until the 1960s, gains in life expectancy were cited as the leading indicator of improving population health. Life expectancy in the United States is defined by the period life tables produced by NCHS, and the trend toward increasing life expectancy is the first health statistic cited in its recent annual report of health in the United States (National Center for Health Statistics, 2007). The statistic may be reported in a number of ways. Life expectancy at birth is often taken as an overall measure of population health, because it aggregates mortality rates for all ages. Life expectancy may also be reported as conditional on achieving a specific age or for subsets of the population. For example, the period from 1970 to 2006 saw an increase in life expectancy at birth in the United States from 70.8 years to 78.1 years (Arias, 2007). In 2005, life expectancy at age 65 was 18.7 years; at age 75, it was 12.0 years. Female life expectancy at birth in the United States was 80.4, exceeding male life expectancy by 5.2 years (National Center for Health Statistics, 2007). In 2003 the United States ranked 26th in female life expectancy at birth among 37 selected countries and territories ranging from Japan (ranked 1st at 85.3 years) to the Russian Federation (ranked 37th at 71.8 years) (National Center for Health Statistics, 2007).

5.1.3. Morbidity and Health-Related Quality of Life

In the modern era of health care, American society has become as concerned with health-related quality of life as with life expectancy (Linder, 1966). A great deal of modern medical care is directed toward reducing morbidity and increasing functioning, thus improving health-related quality of life. Accordingly, researchers have sought to measure not only changes in length of life affected by medical care, but also changes in morbidity and functioning affecting quality of life.

At this point, no consensus summary measure of an individual's health, as affected by morbidity, has emerged. However, most researchers agree that a general measure of health for an individual at a point in time should reflect physical and mental functioning and account for the degree to which a person is affected by pain. These "within-the-skin" attributes form the core of most measures (Patrick and Erickson, 1993). Sometimes additional measures of social and role functioning are included (Stewart and Ware, 1992), which may be affected by physical and mental functioning (as well as other health factors) but are important enough to be valued and measured in their own right. There is dispute about including "beyond-the-skin" attributes—such as social support and the physical and socioeconomic environment within which an individual lives—the dispute being whether these are actual attributes of health or determinants of health. The inventory of attributes of health-related quality of life devised by the World Health Organization (WHO) is perhaps the most comprehensive in inclusion of beyond-the-skin attributes (World Health Organization Quality of Life Group, 1998a, 1998b).

To measure the impact of medical care on health, it seems reasonable to begin by focusing on measures of within-the-skin attributes of health and functioning. Even within this category, many choices of measures remain. An important consideration is whether to use condition- or organ-specific measures or generic health measures. Because any particular health condition or disease may have very specific effects, condition- or organ-specific measures can be quite detailed and sensitive to small changes in how a certain condition affects an individual. As a result, they are more likely than a generic health measure to be the primary outcome in trials of management for particular diseases.

However, the condition-specific options have major problems as measures of general population health. The first problem is logistical. There are many different conditions, and collecting data using a different index for each condition presents practical problems since all persons would need to be asked to identify all instruments relevant to them and to complete those questionnaires. A second and more important problem is that we do not know how to combine the scores from multiple disease-specific instruments into a single summary score to represent the health of an individual who suffers from several conditions (as is the norm for aging people) nor how to compare, say, an increase on an asthma index for one person to a decrease on a diabetes instrument for another person. In addition to the practical difficulty of collecting the data for such a large panel of measures, which simply may not be feasible, is a fundamental conceptual challenge. Converting condition-specific measures into an overall measure of health or well-being requires that they be comprehensive, or nearly so. Furthermore, it presupposes a way to combine them. That is, it is necessary to assign appropriate weights to each condition-specific measure to convert the group of them into an aggregate measure of health. They cannot simply be added together. Aggregation would require substantial new research, including the development of techniques to validate the aggregate measure.

Researchers interested in summary generic measures have instead concentrated on developing measures that cover the major domains of health but do not provide a great deal of detail relevant to any one health condition. The past

HUI3 defines health on 8 attributes (vision, hearing, speech, ambulation, dexterity, emotion, cognition, and pain), each having 5 or 6 levels and jointly describing 972,000 unique health states (Feeny et al., 2002).

Both HUI2 and HUI3 scoring functions have health states scored less than 0 (dead). HUI2 scores range from –0.03 to 1.0; HUI3 scores range from –0.36 to 1.0.

QWB-SA

Permission to use the QWB-SA can be obtained free of charge from the University of California, San Diego, Health Services Research Center (see http://medicine.ucsd.edu/fpm/hoap/index.html). Usually self-administered using a two-sided optical scan form, the QWB-SA assesses health over the past 3 days. It combines three domains of functioning (mobility, physical activity, and social activity) with a lengthy list of symptoms and health problems, each assigned a weight, using an algorithm that yields a single summary score (Kaplan, Sieber, and Ganiats, 1997; Andresen, Rothenberg, and Kaplan, 1998) based on the presence or absence of activities and symptoms on each of the past 3 days. The final score is the average of the three single-day scores. The original QWB algorithm was developed using visual analog scale (VAS) ratings of health state descriptions by a community sample of adults in the San Diego area. The QWB-SA algorithm is conceptually similar to that of the original QWB but was derived from ratings by a convenience sample of people in family medicine clinics around San Diego; VAS were used to rate domain levels and some case descriptions formed from special combinations of domains in a multiattribute utility elicitation process. Excluding dead (0), the minimum possible QWB-SA score is 0.09 and the maximum is 1.0.

SF-6D and SF-36v2

License to administer the SF-36v2 must be purchased from its vendor (see http://www.sf-36.org/). SF-36v2 refers to several time frames. One question asks for self-rated health "in general." Some questions ask how much one's health "now limits" doing certain activities. Other questions refer to the "past four weeks." The SF-6D is computed from a subset of 11 of the 36 questions in the proprietary questionnaire. While the SF-36v2 yields a health profile summary using 8 domains, the SF-6D has reduced this to 6 domains (physical function, role limitation, social function, pain, mental health, and vitality), each comprised of 5 to 6 levels and jointly defining about 18,000 health states (Brazier, Roberts, and Deverill, 2002). Scoring was derived from standard gamble assessments by a population sample from the United Kingdom. The SF-6D scoring algorithm is distributed by the SF-36v2 vendor. The scoring algorithm produces scores ranging from 0.30 to 1.0.

TABLE 5-1 Pros (+) and Cons (–) of the Various Indexes

Attribute	EQ-5D	HUI2/3	QWB-SA	SF-6D	HALex
Nonproprietary	+	–	+	–	+
Low Response Burden	+	+	–	+	+
U.S. Population-Based Weights for Scoring	+	–	~	–	~
Descriptive System Detail	–	+	+	~	–
Applies to Individuals	+	+	+	+	+

NOTE: A tilde represents equivocal, neither strength or weakness.

The SF-6D may also be computed using the RAND-36 questionnaire, which is equivalent to the SF-36 version 1, and available without charge (see http://www.rand.org/health/surveys_tools/mos/mos_core_36item.html). A license to use the SF-6D may be obtained from its developers free of charge for non-commercial applications and for government agencies (see http://www.shef.ac.uk/scharr/sections/heds/mvh/sf-6d).

HALex

No permission is needed to use the HALex. It is the only summary index available that uses data directly from the National Health Interview Survey (NHIS) instead of its own free-standing system. It is used to track years of healthy life in Healthy People 2000 and 2010. HALex questions refer to "your health in general." It consists of 2 domains, 6 levels of activity limitation (ranging from "no limitations" to "unable to perform activities of daily living"), and 5 levels of self-reported health ("excellent," "very good," "good," "poor," and "fair"), jointly defining 30 health states. This is the only one of the six indexes to use self-rated health to describe health states. Appendix 1 of Erickson (1998) abstracts from NHIS the questions used to rate a person in the activity limitation domain. The self-reported health question may be administered alongside these questions to complete an ad hoc questionnaire for the HALex.

The scoring algorithm was developed ad hoc without actual preference survey data using correspondence analysis to the HUI Mark 1. The worst of the 30 health states is scored 0.10, and the best is scored 1.0. Table 5-1 summarizes the pros and cons of the various indexes for use in a health accounts data system. No one measure is uniformly best, and each has strengths and weaknesses.

5.2. QALY AND QUALITY-ADJUSTED LIFE EXPECTANCY AS A SUMMARY MEASURE OF CURRENT HEALTH

The measures of health-related quality-of-life described in the previous section may be used to derive QALY compatible estimates. A QALY is a summary measure of health—based on subjective quantification of illness—that includes both morbidity and mortality. A year in perfect health is equal to 1.0 QALY. The value of a year in ill health is discounted to reflect the relative utility of the ill state versus perfect health; for example, a year bedridden may be valued at 0.5 QALY. In cost-effectiveness analysis of health care interventions, QALYs are now the standard metric for health impacts (Gold, 1996). These impacts are calculated for both individuals getting new treatments and populations with some changes in their health inputs.

A second acronym—QALE (for quality-adjusted life expectancy)—is used in the population health literature as a summary measure of current health status. QALE is life expectancy adjusted for the quality of surviving years and so is measured in QALYs. QALE is by definition less than life expectancy computed in unadjusted years. The discrepancy between QALE and unadjusted life expectancy reflects the relative perceived desirability people place on living a given length of time with morbidity versus living that time in perfect health. In a life-table representation of population health, QALEs are reported undiscounted for time (i.e., just as life expectancies in a population actuarial table are undiscounted).

QALEs and the methods described below to compute them have several good properties. First, they are independent of the age composition of the population. Other measures, such as crude death rates, disease prevalence, or restricted activity days, are highly dependent on age and must be stratified or standardized for many comparisons. Life-table methods are a natural method of standardization that do not require any particular population (such as the U.S. population in 2000) to be chosen. Second, with no additional work, the tables that compute QALE at birth can be used to compute QALE and nonquality-adjusted life expectancy for any age group (e.g., 65-year-old life expectancy as recommended by status of health indicators 2008). Ignoring the health adjustments, these methods compute classic measures of population health, such as life expectancy at birth, to compare with historical data from the United States and other countries. However, using this type of actuarial QALE as a descriptive summary of population health requires cross-sectional surveys of HRQoL, as discussed in the next section.

5.2.1. Cross-Sectional Surveys of HRQoL

Medical care may affect life expectancy, HRQoL, or both. For the purpose of cost-effectiveness analysis of medical interventions, these two are generally combined into one QALY measure (Gold, 1996). At the individual person level, generic health over time may be represented as a function of HRQoL over time, $q(t)$. Once one has observed the individual's HRQoL over time from t_0 to t_1, one can compute the QALYs accrued by the individual between time t_0 and t_1 as the integral $\int_{t_0}^{t_1} q(x)dx$, where the function q is empirically defined by the observations.

More often, empirically defined QALYs are computed by weighting time intervals lived, such as 1-year intervals, by an observed or estimated HRQoL average for the interval, then summing products of HRQoL and interval length across intervals. In this fashion, consider q_a to be the average HRQoL for a person in the year interval from a to $a + 1$. Let ${}_t p_x$ be the probability that a person age x will survive to age $x + t$. The empirically QALE conditioned on current age being a, will be $QALE_a = \sum_{t=1}^{\infty} (q_{a+t-1})({}_t p_a)$. This computation is a variation on standard life-table calculation of age-specific life expectancy where each year of life is weighted by age-specific HRQoL. This method is widely attributed to Sullivan (1971).

Rosenberg, Fryback, and Lawrence (1999) demonstrate this technique using q_a values measured with the QWB index estimated in a community population and combine these with life-table survival probabilities to compute $QALE_a$ for males and females ages 55 and 65. Others have used a binary-valued 0,1 HRQoL function giving disability an HRQoL of 0 to calculate disability-free life expectancy (Molla, Wagener, and Madans, 2001). A similar method that allows for a few states between life and death is multistate life tables (Cai et al., 2003).

These techniques can estimate QALE for a health account. As with ordinary life expectancy, QALE measures do not predict future health, but instead summarize health in the current year. Two inputs are needed: (1) a life table describing mortality experience in the population and (2) average HRQoL at each year of age in a population. Age-specific death rates from NCHS would be needed as the first input. Data from population surveys using any of the HRQoL indexes described above would suffice for the second. In the next section we list existing surveys using one or more of these measures. Table 5-2 shows how the necessary life-table calculations would be structured.

Data for column 1 come from NCHS vital statistics. They calculate the age-specific death rates from data from the decennial census on midyear population of each age, together with their collected deaths by age. Column 2 shows how many people of each age remain alive with these death rates at each age, assuming an initial hypothetical cohort of 1,000 births. Column 3 is new: it would require a population survey of HRQoL. Columns 4-9 are computed quantities using standard life-table techniques augmented for HRQoL weighting, as described above. Column 4 is the product of population × (1 – death rate × fraction of year lost to death). The fraction of years lost to death is usually very close to one-half except for infants because, on average, people die half way through the year. The row corresponding to 100, the largest age in the table, is special, as it covers more than 1 year: the death rate is 100 percent and years lived means expected future years for those exactly 100 years old. Column 5 is the years lived at that age × average HRQoL at that age, so it equals the QALYs lived by the remaining hypothetical population at that age. Columns 6 and 7 are added from the bottom to get cumulative health in years and in QALYs from each age to death in the hypothetical cohort. Finally, columns 8 and 9 divide the remaining years by column 2, the number of people of that age alive, to get life expectancy and QALE.

TABLE 5-2 Life-Table Calculation Framework

	1	2	3	4	5	6	7	8	9
Age Interval	Probability of Dying During Age Interval	Hypothetical Cohort at Start of Interval	Average Age-Specific HRQoL	Population Life-Years Lived During the Age Interval	Populaton QALYs Lived During the Age Interval	Population Life-Years Lived from This Age to Death	Population QALYs Lived from This Age to Death	Life Expectancy at Beginning of Age Interval	QALE at Beginning of Age Interval
0-1 Years	.01	1000	.99	997	997 × .99				
1-2 years		990							
2-3									
...									
100 Years and Over									

These tables also can be used to calculate other generic measures that have been collected for years in many countries, such as infant mortality and (nonquality-adjusted) life expectancy from birth and at other ages, such as 21 and 65. The national health account would need to do so to facilitate historical and cross-country comparisons, although we expect other Western countries to begin calculating and reporting QALE also.

Although restricted activity presumably is reflected in HRQoL, there might be some interest in these numbers and trends as well. One might use life-table methods to standardize other age-dependent measures, such as restricted activity days, calculating the expected lifetime-restricted activity days for a period cohort with death rates and restricted activity days by age as in the current year, but it seems more natural just to report the actual number of restricted activity days, perhaps stratified into large age groups such as children and adults over and under age 65.

Several cross-sectional surveys of HRQoL currently exist; Box 5-1 is a list of data sources. In Table 5-2, new data in column 3 would be the result of periodic HRQoL surveys of the population. Several one-time national data sets and at least two continuing periodic surveys collect one or more of the HRQoL indexes described earlier. However, without augmentation, none of these is entirely sufficient for an ongoing and detailed health account computation of QALE. Three one-time surveys have collected systematic HRQoL data. Although these studies can be used to estimate age-specific HRQoL of community living adults in the United States, they all miss persons younger than age 18, institutionalized persons, and persons living in the community but unable to respond to a survey for physical or cognitive reasons. The Joint Canada/United States Survey of Health was conducted in English, French, and Spanish. The National Health Measurement Study was conducted in English, and the U.S. Valuation of the EQ-5D was conducted in English and Spanish. Hanmer, Hays, and Fryback (2007) discuss similarities and differences of these surveys and implications for HRQoL estimates.

Two ongoing surveys of the U.S. adult population collect information for one or more of the HRQoL indexes. Other studies of note that have included HRQoL indexes include the Health and Retirement Study (HRS, see http://hrsonline.isr.umich.edu/), which administered HUI3 in 2000 as a module for approximately one-twelfth of the full HRS sample, or about 1,600 individuals. The Centers for Medicare & Medicaid Services are required by law to survey a sample of 1,000 Medicare recipients from each participating Medicare Advantage plan. The resulting sample is nearly 100,000 persons per year and has been ongoing since 1998. This survey was formerly known as the Health of Seniors study, but in 1999 was renamed the Medicare Health Outcomes Survey (HOS, see http://www.cms.hhs.gov/hos/ and http://www.hosonline.org/). HOS included the SF-36 version 1 questionnaire through 2007; this questionnaire serves for calculating SF-6D HRQoL scores. In 2008, the HOS plans to change the questionnaire to a version of SF-36 developed for the U.S. Department of Veterans Affairs (Kazis et al.,

BOX 5-1
Cross-Sectional Surveys of Health-Related Quality of Life

One-Time Surveys

- The Joint Canada/United States Survey of Health was conducted in 2002-2003 by the U.S. National Center for Health Statistics and Statistics Canada. Approximately 3,500 Canadian and 5,200 U.S. residents ages 18 and older living in private dwellings were surveyed by telephone using the same questionnaire and methods in both countries. Among many health-related data elements, the survey collected the Health Utilities Index Mark 3 (HUI3). In addition, participants were asked to self-rate their health on a five-category scale from excellent to poor. See: http://www.cdc.gov/nchs/about/major/nhis/jcush_mainpage.htm.
- The National Health Measurement Study (NHMS) (Fryback et al., 2007) used telephone interviews of 3,844 U.S. community-living adults ages 35-89 to collect all 6 of the indexes: EQ-5D, HUI2/HUI3, QWB-SA, SF-6D, and HALex. The NHMS oversampled persons ages 65-89 and telephone exchanges with an above-average density of African American households.
- The U.S. Valuation of the EQ-5D (Shaw, Johnson, and Coons, 2005) was a household face-to-face survey of approximately 4,000 adults ages 18 and older. Participants, in addition to supplying valuations used to create U.S. weights for EQ-5D, self-rated themselves using the EQ-5D and HUI2/HUI3 indexes and the five-category self-rating scale.

Ongoing Surveys

- The National Health Interview Survey (NHIS) celebrated 50 years of health data collection in 2007. The HALex was constructed post hoc to act as a summary index of HRQoL using NHIS data (Erickson et al., 1989; Erickson, Wilson, and Shannon, 1995; Erickson, 1998). See: http://www.cdc.gov/nchs/nhis.htm.
- The Medical Expenditure Panel Survey (MEPS), household component, is a large-scale survey of families and individuals across the United States. It is drawn as a subset of NHIS participants as a sample frame. In 2000 and 2002 MEPS included a mailed-out self-administered questionnaire targeted to persons ages 18 and older. The questionnaire administered the EQ-5D and another health status survey, the SF-12 version 1. From the SF-12 it is possible to compute an SF-6D score (Brazier and Roberts, 2004). After 2002, only the SF-12 has been administered. See: http://www.meps.ahrq.gov/mepsweb/.

2004a, 2004b, 2004c). Whether SF-6D scores can be calculated after this change is not known at this time.

None of the surveys mentioned above consistently includes actual medical examination of participants but, beginning two waves ago, the HRS has been collecting blood and some anthropometric measures. The ongoing National Health and Nutrition Examination Survey (NHANES, see http://www.cdc.gov/nchs/nhanes.htm) does include a medical examination and testing of participants. To assist in modeling relationships of proven medical conditions (versus self-reported ones) for HRQoL, it would be highly desirable for NHANES to include at least one of the HRQoL indexes in its protocol. More broadly, given that multiple surveys (each with a somewhat different population scope) are required to adequately measure health across the population broadly, there should be some effort by the statistical agencies to pick a common quality of life instrument to use in the different surveys.

Recommendation 5.1: A committee of members from agencies responsible for collecting population health data (Agency for Healthcare Research and Quality, National Center for Health Statistics, Census Bureau, etc.) should be charged with identifying and putting in place a single standard population health measurement tool (or set of tools) to use in a wide range of surveys. The best instrument, which is situation specific, may simply be the one that can be added to enough surveys collected over time so that most of the population is covered.

Ideally, agencies would collaborate and choose one instrument that can be followed over time, for the purpose of having at least one comparable measure for different years, but others could be considered as well. For example, it would be very useful if a generic quality of life measure were added to the Medicare Current Beneficiary Survey (MCBS). The MEPS and MCBS should pick at least one instrument in common that both will consistently use over time. Although the choice of instrument should be left to the agencies, the system should strive for consistency of instrument use over time. The main reason for standardization is that chaining from one instrument to another is problematic. If a change does occur, an overlap period is needed for the transition.

5.2.2. Disease-Specific Measures of Health

In addition to generic measures of health, the national health accounts should also contain information on a set of specific diseases. WHO, for example, collects data on the incidence, prevalence, patterns of treatment, and mortality from tuberculosis for almost all countries in the world, and these are reported annually (World Health Organization, 2008). In the United States, separate data collection efforts are ongoing for cancer, HIV/AIDS, end-stage renal disease, and other diseases, the results of which can be used by researchers and policy makers. The databases may include information on stage at diagnosis, treatments, disease severity, and mortality. Is it reasonable to try to include such data on all major diseases in the health accounts?

Disease-specific data are considered by many to be the cornerstones of health research as they can provide the basis for learning, for example, if the incidence or prevalence of Alzheimer's, HIV/AIDS, or lung cancer is going up or down; how limiting is arthritis or asthma; or how many people are dying from food poisoning or heart attacks. Disease-specific data may fit more closely with the intermediate outcome of expenditures organized by disease-specific episodes of treatment that are used to estimate improvement per expenditure. They are vital in planning policies and research on these diseases and in evaluating current programs.

However, as a general population-wide measure of health, disease-specific data have several problems. To be comprehensive, an enormous number of diseases would need to be covered, so the price tag would be high. Aggregation into a single or a few measures seems very difficult. How would one add up cases of asthma and breast cancer, or even cases of breast cancers of different severity? How would one treat cases of cancer in remission or aggregate cases with different stages? Also, each disease-specific health unit is on its own scale. Researchers are now in the early stages of trying to estimate the QALY impact for specific diseases on future health and survival. In one of these initial studies, Eggelston et al. (2009) measured the net economic value of improvements in health status (to an admittedly homogenous population), using a QALY metric defined in terms of 10-year cardiovascular risk, from spending on care for patients with type 2 diabetes.[3]

Another issue is the overlap of diseases, which raises the problem of allocating the total health decrement to each that is present. It is even difficult to split up deaths in this way—given that multiple possible causes of death are commonplace, information is currently not precise enough to confidently estimate death trends by disease. It is also useful when these measures include cognitive impairment. This is a large problem for coverage of the elderly and also for populations outside the United States—for example, deaths of children in developing countries from malnutrition.

Recommendation 5.2: Recognizing the difficulties in estimating the incidence and prevalence of disease, the National Center for Health Statistics should commission research on selecting and specifying a set of important acute and chronic diseases and feasible methods for estimating acute incidence and chronic prevalence that might be part of a national health data system. These

[3]Assuming that "one life-year is worth $200,000 and accounting for changes in modifiable cardiovascular risk," the authors found that "the net value of changes in health care for patients with type 2 diabetes was $10,911 per patient (95% CI, is $8,480 to $33,402) between 1997 and 2005."

counts would be a supplement to the systematic data systems to measure health-related quality of life using measures that transcend single diseases (Recommendation 5.1).

Such estimates should prove useful for looking at the impact of trends in screening, expenditures per case, and deaths.

5.2.3. QALYs and Disability-Adjusted Life Years

A possible alternative to the QALY approach for measuring population health stock would be to base it instead on the disability-adjusted life year (DALY) methodology (Murray, 2002). These measures were developed to help WHO and others working in the developing world understand the burden of (generally infectious) diseases, plan and evaluate policies to reduce the impact of those diseases, and assess health using data that might be available in those countries.

The DALY approach to measuring health is based on disease incidence and prevalence—a fixed burden per person is computed based on the methodology, and then total burden is counted by multiplying this times the number of people affected. Early versions of the methodology used expert opinion to set the burden per person of morbidity, but the DALY weights and methods overall have been fluid in the past several years in response to criticisms, and they are now much closer to a QALY. One minor difference is that the DALY method assesses a health gap: the difference between a person's life and a life at good health to some maximal age, whereas QALY methods typically say how much life is achieved. If the methods for assessing HRQoL were identical, there would be the identity that the QALYs (at less than maximal age) + DALY gap = maximal life span. The developing world's focus on DALYs may be partly driven by the enormous data gaps that exist, and it makes sense conceptually for those trying to maximize the reductions in disease-based disability and death (burden of disease) for a given cost outlay. However, they are not sufficient for that purpose, because estimates of the effect of any program on each DALY are also necessary.[4]

The QALY method is much more widely used than are DALY methods in the United States and among the developed countries in the Organisation for Economic Co-operation and Development. In these countries, the major health problems are no longer childhood infectious diseases but the chronic conditions that people accumulate with age and need to manage to preserve their health. Also, the QALY methods have been the subject of extensive scientific investigation and were, in particular, developed for the health questions typically asked on the ongoing surveys. Economic accounts are in the business of organizing data for measuring the value of goods and services generated in the economy, so it makes

[4]There is also dispute about the DALY disease weights. For a good summary of differences between QALYs and DALYs, see Sassi (2006).

sense to think in terms of QALYs (the good that is being generated from medical care). And for a U.S. application, one would want data of sufficient quality to estimate QALE using QALY methods.

All things considered, for the purpose of measuring population health for a national health account, QALY is the most appropriate metric currently available.

Recommendation 5.3: Initially the national health account should focus on quality-adjusted life expectancy measured in qualify-adjusted life years as the best summary measure of health in each year.

Data on important risk factors that impact future health should be collected, but more research is needed before useful calculations can be made of the future life expectancy of the current population.[5] Expected future QALYs will need to be revised as the effect of treatments become known—this is similar to other parts of the NIPAs.

5.3. IS MEASURING CURRENT HEALTH BY QALE ENOUGH?: ADDING RISK FACTORS TO THE DATA SET

The computations of QALE outlined above are not forecasts of future health. They provide a picture of current health in the United States—a snapshot of current health accrual. For the reasons discussed, QALE computed from period life tables is not a forecast of experience of the population in the future. At this point, a national health account should focus more on current health than on predicted future health, mainly because data will be easier to collect, assemble, and understand. Much can be said about particular risk factors in terms of their future impact in a national report, but overall life expectancy of the current population is a stretch.

Data should include generic health in the current year and information on risk factors and disease. Average HRQoL, including death, is appropriate for the current year, and it may be age-adjusted or not to track over time. Currently life expectancy is based on this year's outcomes, not on projections, for the practical reason that they are easy to compute. Population-based QALE is a good metric (and an improvement over life expectancy) for current health stock, although it is data-intensive to compute.

That said, it is important to recognize that many medical, personal, and public health interventions are investments in future health, not just current health.

[5]This recommendation is consistent with a recommendation of an Institute of Medicine panel on valuing health for the purpose of cost-effectiveness analysis (Institute of Medicine, 2006). That panel recommended the use of QALYs when evaluating medical and public health programs that primarily reduce both mortality and morbidity: "Regulatory [cost-effectiveness analyses] that integrate morbidity and mortality impacts in a single effectiveness measure should use the QALY to represent net health effects" (p. 161).

For example, one of the most successful interventions of the past four decades to which a good proportion of current gains in health is attributed is detection and treatment of high blood pressure. The purpose of diagnosing and treating high blood pressure is not to improve current health but to improve future health—indeed, many drugs aimed at preventing future heart disease may even have side effects that reduce quality of life in the short run. Similarly, many environmental policies are aimed at lowering the burden of future chronic disease by reducing exposures to carcinogens or other air pollutants, even when those exposures have no impact on present health (Gold, 1996).

The purchased medical services tracked in the medical expenditures accounts are an important input into the production of future health, but there are many other inputs. The National Research Council report *Beyond the Market* (National Research Council, 2005) lists five other major inputs: (1) medical care services provided without payment; (2) time and effort that individuals invest in their own and relative's health; (3) consumption of nonmedical goods, such as food, tobacco, and alcohol; (4) research and development that may lead to improvements in medical technology; and (5) environment and disease state factors and shocks. In addition, one might add other influences, such as immigration, education, social capital, culture and desires, and genetics. Finally, the success of earlier treatments leads to populations that are older and have a history of disease. These factors aggregated from conception to the present all impact health.

Cost-effectiveness studies of medical technology do need to forecast future health and usually do so through use of mathematical modeling of treatment results and personal risk factors associated with patients. Risk factors are characteristics of individuals or individual behaviors that are associated with future health gains in the population. A comprehensive list of these is beyond the scope of this report. However, an accounting of health gains from current medical investments needs to recognize that reductions in risk factors in the present can improve QALE in the future. We suggest several criteria for deciding which, if any, inputs to future health to collect:

- There should be feasible or actual reasonable national estimates available.
- Solid scientific links should exist between the risk factor and future health.
- The impact on future health should be large enough to make data collection worthwhile.
- Interventions to change the risk factor should be reflected in the expenditures part of the national health account.

By these criteria, data concerning risk factors in the United States today should include information related to:

- smoking and tobacco use;
- physical activity;
- sleep habits;
- obesity (i.e., body mass index, although waist/height may work better as a predictor);
- measures of diet (apart from obesity);
- high blood pressure;
- cholesterol;
- alcohol use, especially when driving;
- education;
- birth weight and prematurity;
- participation in screening (e.g., mammography, colorectal cancer screening);
- vaccination (e.g., childhood, pneumococcal, human papillomavirus);
- oral hygiene;
- preventive interventions (e.g., seat belt use);
- exposure to environmental factors (e.g., violet A and ultraviolet B light); and
- diseases such as diabetes, chronic kidney disease, cardiovascular disease, and cancer, which have a major impact on future as well as present health.

Most of these are collected regularly through the Behavioral Risk Factor Surveillance System (BRFSS) of the Centers for Disease Control and Prevention (CDC). BRFSS is a state-based system of health surveys conducted by telephone. CDC assists states to track health risks in the United States and makes the data widely and easily available. A number of states use the opportunity to make BRFSS a rolling cross-sectional survey; however, CDC publishes the data as annual surveys, allowing tracking of trends over time and comparisons to be made at the state level.

Most (not all) of these risk factors are collected as self-reports. The NHIS asks if a doctor or medical professional has told the person that they have high blood pressure or high cholesterol, and many surveys have copied this format when it is not possible to document with actual physical or laboratory measurement of blood pressure or cholesterol. NHANES makes actual measurements on participants for these data. From NHANES one could add measures of stress and of the incidence and prevalence of such diseases as diabetes and cancer that presage future poor health. It is unlikely that self-reports of biological variables such as cholesterol, cancer, or other lab work will typically yield anything like complete or accurate results. NHANES, together with an algorithm, has been used for adults to define chronic kidney disease in a microsimulation model; it uses a correction factor in which a percentage is taken off for early-stage chronic kidney disease because proteinuria is measured only once, but that correction seems solidly based (Coresh et al., 2007).

In general, however, alternatives based on self-report (e.g., "Has a doctor told you that you have diabetes?") are unlikely to give rise to very good models of future health, which means that the predictions of future health in these cases will be somewhat speculative. The spread of electronic medical records may facilitate data collection on the factors that are usually in such records. Clearly, better markers are needed of current population health and risk factors to be able to project future health better.

Attribution of health trends to changes in inputs is discussed in Chapter 6. In reports on risk factors in the population, rough interpretation of trends using the current epidemiological literature may be applicable. For example, one might say that the observed *y* percent lower hypertension prevalence will lead to *x* percent fewer deaths using the formulas derived from Framingham Heart Study data. One such formula calculates risk of cardiovascular disease based on smoking, diabetes, systolic blood pressure, cholesterol, high-density lipoproteins, gender, and age (D'Agostino et al., 2008).

5.4. HOW CAN HRQoL AND RISK FACTOR DATA BE COLLECTED FOR A NATIONAL ACCOUNT?

There are three problems with currently collected HRQoL data:

1. Sample sizes, although large for most health research, are small for making detailed analyses.
2. National surveys are mainly limited to adults living in the community and miss children as well as institutionalized persons and those living in the community but who are cognitively impaired.
3. Release of national survey data often lags 1-3 years from the time they were collected.

Unlike data for national economic accounts, for which data are often collected and reported quarterly, health data are generally collected and reported much more sparsely. BRFSS data are collected annually; however, the BRFSS website in April 2008 reported trend data only up to 2003. NHIS and NHANES data are available for 2006, with selected estimates recently released from 2007 data. MEPS released 2006 data in phases throughout calendar 2008. Mortality data typically lag 2-3 years. A lag of 1-3 years for data reporting may be sufficient for a satellite health account. However, if the nation wants health accounting anywhere near as current as economic accounts, then the collection of health data needs a paradigm shift. And there may be such a paradigm shift on the way.

As part of the "Roadmap" of the National Institutes of Health (NIH), which is a priority path for medical research, NIH has funded a consortium of researchers to develop a standardized system for collecting patient-reported health outcomes data: the Patient Reported Outcomes Management and Information System

expressed either in years or dollars (Murphy and Topel, 2006). In addition, we discussed briefly how the risk factors that predict future health might be chosen and collected. In the health account, we proposed expressing health outcomes in both natural and monetary units. Monetary valuation of health is difficult: even when the provision of medical care involves prices (although often ones that do not closely reflect cost), some inputs, such as volunteer labor for the chronically ill, do not. Also, nonmedical nonmarket inputs include time invested in one's own health (for example, exercise and sleep) or in a relative's health, and such activities often have additional goals besides future health, which complicates evaluation.

A comprehensive—and at this point admittedly futuristic—health account would also attempt to incorporate topic (4). It would not only identify, quantify, and value the flow of nonmedical health inputs, such as behavior trends (e.g., diet, risk taking, smoking, consumption of alcohol), research and development, and the quality of the environment; it would relate both these and medical inputs to current and future population health. While emphasizing the value in monitoring both inputs and outcomes, we have been largely agnostic on exactly how researchers go about the task of quantifying causal links between medical care, health-enhancing activities, and other inputs to the population's health through disease modeling. This is a difficult area of inquiry, both conceptually and in terms of data requirements, that is being pursued in leading-edge research taking place across many institutions, primarily on a disease-by-disease basis; it is the type of work that BEA will probably never do. That said, results from this research could eventually be used to enhance the usefulness of a national health account that the statistical agencies play a role in constructing. While it is beyond the scope of this study to offer detailed recommendations on this academic research, in this chapter we review some of the ongoing work, offer some general guidance for U.S. efforts going forward, and describe how a national health account would provide a useful centralized data depository from which investigators could draw and into which results may feed.

Much of this chapter is concerned with developing a data system that would allow changes in the population's health (death and impairment) to be linked to changes in spending on medical care and other factors. While the panel is skeptical about how well, at least in the short run, outcomes can be linked to medical expenditures and other factors, we strongly recommend beginning the process of gathering data in a way that improves the ability of researchers and policy makers to draw causal inferences. Creating and pooling electronic health records (discussed more below) would seem to be a prerequisite on which to focus this line of development.

6.1. ATTRIBUTION OF HEALTH EFFECTS TO INPUTS

As discussed in Chapter 2, a major policy issue motivating research on the topic of this report is how to gauge the productivity of the medical care system, which provides one rationale for beginning there. For this purpose, it is important that investigators have data that allow them to attribute population health effects to factors that work separately or interactively with medical care. It is advisable to begin by trying to do the medical part well, but elements included in a broader boundary of health goods, services, and activities become more important when trying to determine causality for outcomes.

There are many ways of linking medical care and other inputs to health. One is through a standard medical trial of certain inputs or interventions (e.g., diet counseling), looking at the effect of treatment on a primary outcome in which the investigator attempts to keep other factors determining the outcome constant by standardizing patients and treatment and using randomization. Another approach is the one discussed in Chapter 3, whereby econometric methods are applied on national data over time on multiple inputs and health outcomes, together with clinical insights, disease modeling, and common sense to figure out what is causing what. Intermediate approaches include epidemiological studies using specialized panel data sets such as Framingham or Surveillance, Epidemiology, and End Results (a registry of the National Cancer Institute). These specialized data sets become more useful when information from patient claims have been linked to them. Despite the problems with trials (e.g., cost, delay, external validity), we certainly support the paradigm but want to open up the possibility of using other approaches for developing data linking medical treatment and other inputs to health outcomes. Improved data on expenditures, prevalence, and death that are classified by disease will be useful for other research projects as well. They may be used for constructing comparisons over time or across countries, regions, or subpopulations in terms of burden.

Whatever methods are pursued, a system for attribution entails more than just collecting data on the multitude of factors that affect population health. Such a data system is merely a tool to help researchers working in this field. While data on high blood pressure and other personal risk factors should be collected and presented, attribution is very difficult for a number of reasons—perhaps one can say that lower hypertension will lead to fewer deaths, but overall mortality is based on many things that occur in the past, present, and future.

The problem is most severe when the objective is to attribute outcomes to services or causes; in many cases, the medical linkages are not known or well understood. For example, in the United States (and many other countries), functional limitations in the elderly will be much more important in the coming years, simply because of the demographic shift. Mobility impairments tend to be the result of multiple medical conditions, some of which are not ordinarily thought of as diseases. Arthritis (most commonly osteoarthritis, but sometimes cartilage problems, rheumatoid arthritis, or other conditions) is usually a component. Poor balance and impaired proprioception may also be important, and these may be the result of strokes or simply "aging," meaning that there is some neurological problem but no known disease that caused it. Projecting the effects of, for example, joint replacement may be difficult because the underlying arthritis is only one of the causes of the mobility impairment. In other words, there is not a one-to-one correspondence between the functional limitation and a disease. So the boundary and attributional issues are likely to become even greater challenges.

Even when the change in a person's health state can be attributed to one condition—say a chronic disease—defining an episode can be difficult. Furthermore, for an elderly population with multiple chronic diseases, the determination of a primary diagnosis for a hospitalization can be somewhat arbitrary. Someone may truly be admitted because of an exacerbation of heart failure, but he or she might have needed hospitalization only because of impairment from other conditions; otherwise, he or she might have been treated on an outpatient basis.

Ideally, the complementary surveys collecting quality of life data (discussed in Chapter 5) could be better coordinated with common survey questions identified to get at nonmedical care inputs measured consistently over time. It would be valuable to assemble aggregate data on all the conceivable determinants of health in some researcher-accessible location. Depending on what kind of method will be used for attribution, very aggregate data may suffice (over time or across countries). The national health account program could begin accumulating data on time use (particularly in preventive activities), consumption trends, other risk factors, behavioral trends, the environment, etc. Even before these data are integrated into a health account, such a data clearinghouse would give researchers attempting to link cause and effect a starting place. Such a data system will be in a constant state of evolution.

With better data or understanding of causes, data components could be expanded incrementally and coordinated, and, indeed, the determinants of a population's health will change over time, in terms of both the set of relevant factors and the impact of each. Environmental factors are notoriously difficult to pin down, since, even if measures like ambient air and water pollutants at a given site over time are available, it is hard to determine an individual's location at all times to derive exposure values. This is also a problem for occupational exposures. It will be a big step to develop more robust data sources on personal factors like blood pressure, total cholesterol, and the glomerular filtration rate; however, their determinants (e.g., diet, air pollution exposure) involve physical or laboratory measurements, and it would be a huge undertaking to add these to surveys. That said, the Health and Retirement Study (HRS) undertakes similar tasks that can serve as a model for learning more about experience with, and the cost of, such additions.

6.2. MEDICAL CARE EXPENDITURES AND HEALTH

The idea of a data system that coordinates information about medical spending, health outcomes, and the population's quality and length of life may be relatively new in the United States, but efforts internationally trace back further.

Statistical agencies in all developed countries—including the United States—produce at least some components of a national health account, since all calculate total medical spending, and many perform cost-of-illness analyses (Heijink et al., 2008). Several countries have gone further, measuring health trends in addition to expenditures with the objective of facilitating disease-by-disease comparisons. In this section, we do not add to the extensive discussion from Chapters 1-4 of the medical care input to health; we only identify a sampling of efforts that have been made by government agencies to begin establishing linkages between medical expenditures and health.

Statistics Canada, the country's statistical agency, has been experimenting with health measurement since the early 1990s. The agency has adopted the Health Utilities Index Mark 3 (HUI3) health assessment tool (see Chapter 5) and has incorporated it as a permanent component in its National Population Health Survey (Statistics Canada, 2007). Using these surveys, the government reports health-adjusted life expectancy at birth and at age 65, stratified by gender, province, and income group. The HUI3 has also been used by the agency to estimate trends in the health impact of various diseases. At the same time, the Canadian Institute for Health Information maintains the National Health Expenditure database, which tracks annual medical spending in the country (Canadian Institute for Health Information, 2006). Expenditure estimates are reported separately for over 40 disease categories, 5 payer sources, and for each province. These two agencies have joined together to publish national estimates of the economic burden of illness for the years 1987, 1993, and 1998 (Health Canada, 1998). For each of 20 disease categories, these studies report direct costs (hospital, physician, drugs, research, institutional, other) and indirect costs (premature mortality, long-term disability, and short-term disability), stratified by province, age, and gender. Over the decade, these studies have been enhanced by methodological refinements and by collection of more detailed data.

The Australian Institute of Health and Welfare divides medical spending for the country into 176 disease categories in such a way that accounts for 94 percent of medical spending. Expenditure data are available by age, sex, and service category (though not for every year). The accounts have been linked to population health data from Australia's Burden of Disease and Injury Study (Mathers, Vos, and Stevenson, 1999). Investigators from this study used methods similar to those of the Global Burden of Disease Study (Murray and Lopez, 2006) to estimate the health of the Australian population in 1996 and 2003. Efforts are now under way to estimate the returns to medical spending for each disease category.

The motivation for much of this work is to improve the productivity of health care systems—that is, to:

1. Produce outputs of health services with a minimum of real resources (technical efficiency) at each level of care, while also minimizing the (relative) costs of inputs (cost efficiency).

2. Provide a mix of care that maximizes impact on health outcomes at a minimum input cost (cost effectiveness).
3. Set overall resources for health care consistent with achieving wider goals of social welfare and to allocate services across individuals at levels that make the best use of these resources (efficiency and equity in resource allocation).

Technical or cost efficiency can be high if a given set of medical activities—or outputs—are produced with a limited amount of inputs (see Figure 6-1). However, if the impact on health status is limited, little social value is obtained from these outputs. This is why the medical community focuses more on cost-effectiveness as it provides a measure of the actual health returns to spending (Gold, 1996).

There are a number of challenges to establishing health outcomes as the standard in the actual practice of system evaluation. First, for some interventions, it can take a long period of time before they have any significant effect on health. Second, there are not always demonstrated links between interventions and health. For example, available data may show an impact only on behaviors that affect health. Alternatively, the data may show no impact at all, and frequently there are simply no data on the effects of an intervention. Finally, when data do exist, the data collection and surveillance systems may not provide the level of detail necessary to measure desired changes in health. The goal, then, is to strike an appropriate balance between intermediate output measures and longer term health outcomes. The challenge is in defining those outputs and in gaining access to sufficient data to measure both prices and quantities.

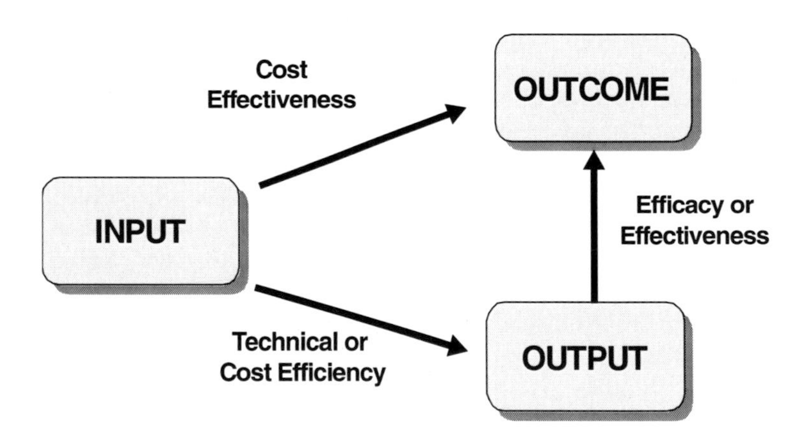

FIGURE 6-1 From inputs to outcomes.
SOURCE: Adapted from Joumard and Häkkinen (2007, p. 12).

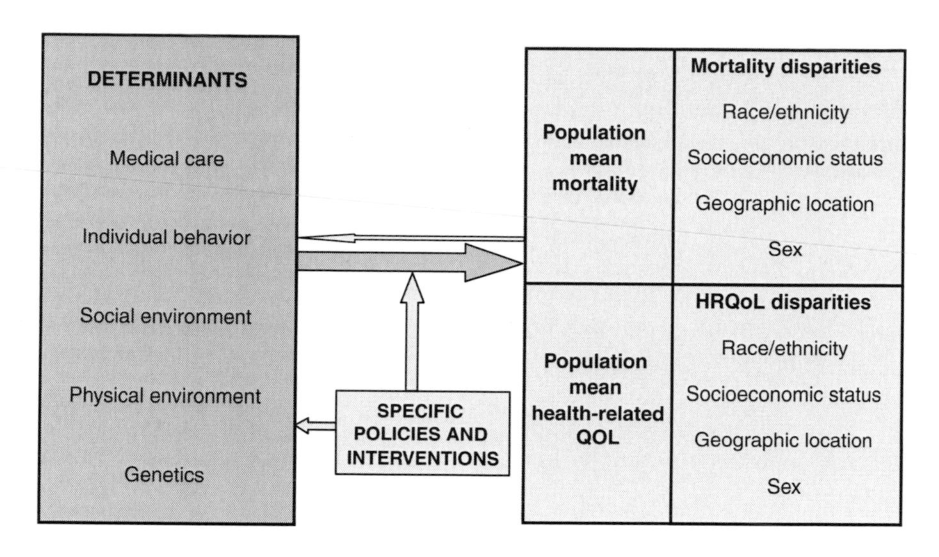

FIGURE 6-2 A schematic framework for population health planning.
NOTES: HRQoL = health-related quality of life. QOL = quality of life.
SOURCE: Kindig, Asada, and Booskie (2008). Reprinted with permission.

6.3. NONMEDICAL AND NONMARKET INPUTS TO HEALTH

Chapter 2 lays out, in general terms, the structure of an accounting system that includes, on one side, data on the inputs to health and, on the other, data on the output, defined as population health. *Beyond the Market* (National Research Council, 2005) describes a similar structure. Figure 6-2 summarizes the basic elements of this relationship. Generating the data and specific accounting structure needed to quantify these relationships is a much more difficult task. Complicating the goal of establishing links between inputs to health and health itself (and even just figuring out which kinds of data to collect to inform the task) is that, as shown in the figure, health is a function of much more than just medical care.[1]

Population health, measured in terms of life expectancy or more subtle quality of life metrics (discussed in Chapter 5), is mediated largely by such factors as personal behaviors (e.g., sedentary lifestyle, smoking), environmental exposures, and public health measures that transpire outside the medical care setting. While health policy gives some attention to public health issues, it deals little with the social context of life, which can exert profound influences on health (Woolf, 2009). Dramatic disparities affect poor and minority populations, who endure poorer health and on average die younger than more affluent groups.

[1]Research on this well-established relationship has its antecedents in research by McKeown (1976), Fogel (1986), National Research Council (1993), and may others, relating determinants of health and life expectancy.

While social determinants such as education, income, and race are clearly interrelated, they exert independent effects on health as well (Robert Wood Johnson Foundation, 2008). A full explanation of changes in population health requires information on multiple aspects of the social context. Furthermore, if health consequences arising from social policies originating outside the health care sector are to be monitored, data collection in less traditional settings—such as schools and community recreation centers—will eventually be needed.

Solving the nation's most pressing health care problems, then, will require a greater understanding of the full range of the factors that determine health and of their complex interrelationships. It is increasingly recognized that the most urgent public health challenges cannot be adequately addressed within a single discipline but instead require a more comprehensive approach. Kindig sets forth a schematic framework for population health planning—on which Figure 6-2 is based—that provides a preview of both data collection and the cross-disciplinary expertise needed for developing a health account (Kindig, Asada, and Booske, 2008).

Perhaps the most important message conveyed by this framework is that population averages can be deceiving. Mean mortality or health-related quality of life measures mask real disparities in outcomes—disparities that can be identified only with adequate data (compiled at a sufficiently disaggregated level) on the social determinants of health. Kindig divides these determinants into five categories based on the Evans-Stoddart model (1990)—(1) medical care, (2) individual behavior, (3) social environment, (4) physical environment, and (5) genetics—to which one might add age, gender, and medical history; this top-level organizational structure may be a good place to start when specifying needs for a national health data system. While some population surveys in the United States (such as National Health and Nutrition Examination Survey [NHANES] or HRS) are a good source of nonmedical health data, ultimately data on determinants of health will need to come from multiple sources as there are many other variables—safe sex practices, occupational and geographic exposures, the physical environment, and others (many of which can be very difficult to measure)—that are not covered in any single survey.

6.3.1. Valuing Informal Care and Other Time Costs

As emphasized above, a comprehensive health account requires tracking the full range of factors that affect health, even if they do not entail market transactions. Indeed, not even all medical service inputs are reimbursed. Many of the quantitatively significant influences on health—some relating to medical care and some not—can be linked to the way in which the population spends its time. A population that spends time actively, in physical exercise, for example, will be healthier than an otherwise similar sedentary one. Another example of a nonmarket cost is waiting time—time spent in physicians' offices or hospitals waiting for care to be provided.

The most quantitatively significant of these services is care provided by relatives. The amount of quality-adjusted time that people spend caring for the ill should be positively related to the health outcomes of the ill person (although, frequently, it is negatively related to the health outcomes of the caregiver). Unlike formal home health care, unpaid caregiving is not included in the National Health Expenditure Accounts (NHEAs). As a result, NHEAs understate total resource use in the care of the ill (or the very old or very young, for that matter, which may also affect health). Furthermore, there is a bias in estimates of the growth in resource use over time, depending on whether informal care is rising or falling relative to market care. The issue is analogous to the treatment of home production in the National Income and Product Accounts. Market-purchased services (paying a laundromat, going to a restaurant) are counted as part of gross domestic product (GDP); home production (doing the laundry at home, cooking at home) is not. As a higher percentage of women have entered formal labor markets over recent decades, and more services previously provided at home are now purchased, estimates of GDP growth could in theory exceed actual increases in valued economic activity due to the displacement of home production.

Almost all analysts of national income accounting—including those who produced the report *Beyond the Market* (National Research Council, 2005)—argue that nonmarket activities that are very close substitutes for market counterparts ought to be included in at least some version of GDP. Indeed, the same study states (Recommendation 6.3) that, ideally, estimates of the value of nonmarket medical care inputs, including time use, ought to be included in national health accounts. Time spent providing health-related services would be valued based on a replacement labor-cost approach; time spent in activities that improve or maintain one's own health would be valued using an opportunity-cost approach (see National Research Council, 2005, pp. 127-130, for a full discussion of this rationale). These estimates may be small initially because of inadequate data on many aspects of time use, but this will challenge statistical agencies and researchers to call for improved data. Indeed, one reason for including these inputs in an account is that the value of time resources expended could be quite large relative to the value of market-provided services, particularly in such areas as elder and child care (LaPlante et al., 2002). Rules will have to be established dictating exactly what kind of time expenditures should be included.

Work to quantify the amount and value of time inputs to health will be constrained for the foreseeable future by data availability, although the survey options are growing. For example, data on various components of informal care provision are available in currently conducted surveys. The HRS asks respondents to report activities of daily living—such as ambulating (walking), transferring (getting up from a chair), dressing, eating and drinking, performing personal hygiene, taking medication—or instrumental activities of daily living—such as driving, preparing meals, doing housework, shopping, managing finances and medications, and using the telephone.

6.4. DISEASE MODELING

Developing national health data in a way that informs resource allocation policies by allowing researchers to causally link medical care and other spending to health improvements is a great methodological (and data) challenge. We have argued that attribution of health effects to spending is most straightforward in a disease or episode context, which suggests a certain modeling approach. Detailed disease models—which estimate interactions among risk factors, specific diseases, and health outcomes—allow analysts to infer the value of medical care at that level.[4] In principle, one can then aggregate across diseases to estimate the productivity of medical care as a whole. The more detailed the disease models are, the greater is their potential to help identify clear targets for more nuanced policy interventions. A full accounting is a long way off at this point because only scant data exist that would allow modeling the joint distribution of environmental and other factors. Nonetheless, the goal should be to develop detailed disease models that relate health inputs to outputs and that will allow researchers to infer the value of medical care at the disease level so that, when aggregated, it may be possible to estimate the productivity of medical care as a whole (Rosen and Cutler, 2007).

Due to the historically slow accumulation of this research, integrating health expenditure and outcome data is a long-term project. The time-consuming disease modeling required is particularly difficult for such pathologies as hypertension that have future consequences that differ from today's. The missing link is how to attribute deaths and changes in health in the presence of multiple conditions and sequenced conditions (e.g., depression leading to obesity, leading to heart disease, leading to death). This kind of scientific endeavor is beyond the scope of traditional economic accounting as these relationships must be established in the literature on a case-by-case basis (e.g., Cutler et al., 2001).

In other disciplines (beyond economics and economic accounting) such as decision sciences and industrial engineering, an extensive history of disease modeling has developed. A comprehensive catalogue of such models spanning over 25 years of the medical literature is maintained by a team of investigators at Tufts Medical School as an Internet-based resource (see http://www.tufts-nemc.org/cearegistry/data/default.asp). These range from simple explanatory models of a single therapy for a single disease (for example, antibiotics for childhood ear infections) to broad policy models that consider several services simultaneously

[4]In contrast, in the absence of a clear relationship between outcome gains and cost increases, highly aggregated cost-effectiveness ratios may be misleading. Studies by Fisher, Skinner, and others suggest large regional differences in spending that are relatively uncorrelated to health status (Fisher et al., 2003; Skinner, Fisher, and Wennberg, 2005). Although causality in this relationship can run in both directions, residing in areas with markedly higher health care spending does not appear to confer better health outcomes. While aggregate trends are informative, more detailed models are needed to better understand (and convey to policy makers) how best to maximize the productivity of health care spending.

(e.g., the prevention and management of coronary heart disease). However, there is no consistent set of conventions that would allow these models to be merged together to provide a picture of the health care sector as a whole.

The starting point, one that allows for the possibility of using different output metrics, is the specification of a general model of the health care portion of health attribution. For a given level of quality, health care outputs are produced from a set of health care inputs. For any given combination of inputs, however, variation in output can be expected due to differences in patient characteristics (such as comorbid conditions and disease severity) or case mix. Hence, we incorporate case mix into the production function, which can be expressed formally as:

$$\text{outputs} = f\,(\text{inputs, quality, case mix}).$$

Case-mix measures use administrative data to classify patients, encounters, or other units of output into homogeneous groups with similar expected health services utilization according to characteristics, such as diagnoses, disease severity, demographics, or a combination of these (Smith and Weiner, 1994; Berlowitz, Rosen, and Moskowitz, 1995). The basic principles underlying the development of these groupings—and their usefulness to BEA—are that they should be clinically meaningful, resource homogeneous, limited to a manageable number of groups, and derived from regularly collected data. There are several case-mix measures on the market today, many made for different purposes in different populations (Rosen, 2001; Iezzoni, 2003; McGlynn, 2008). The two main categories of efficiency measures are episode-based and person-based (or population-based). These approaches, as they apply to creating expenditure accounting categories, are described in Chapter 3.

For modeling outcomes associated with expenditures on specific disease treatments, ideally, the episode-of-care approach should be defined by clinical diagnoses (e.g., back pain) and not by the procedures used to treat those diagnoses—especially if there is any question of appropriateness of use. Removing procedure codes from the definition of episodes would likely resolve the issue but the commercial grouper software (discussed in Chapter 3) varies in its reliance on these codes.

A potential drawback to the episode-of-care approach is that it may inadvertently make primary care appear inefficient. For example, chronic diseases such as congestive heart failure and diabetes tend to be lifelong, requiring long-term therapy for patients. Individuals with these chronic diseases may have flare-ups requiring hospitalization (and triggering the start of an acute episode) with no fundamental change in their underlying chronic condition. For these patients, paying for each hospitalization on an episode-of-care basis may help control the costs of each hospitalization, but it does nothing to control the number of episodes (hospitalizations) that the person experiences. While good primary care can prevent acute exacerbations and long-term complications of chronic diseases (Starfield, Shi, and Macinko, 2005), an episode-of-care output measure will not capture these

6.4.2. Desired Characteristics for an Integrated Model

A number of features and capabilities desirable in an integrated health policy model can be identified. First, the model should relate to the broadest possible population (i.e., not just subpopulations for which there are more data). Furthermore, the model should serve as a platform to inform a broad range of policy decisions, capture health outcomes and costs at the disease level and over time, provide a sense of the level of uncertainty around particular findings, and be able to incorporate new information on an ongoing basis.

While this may sound straightforward, in practice it is quite complex. The overarching system has multiple complicated, interacting parts. What is needed is an approach capable of capturing the essential interrelationships as well as the feedback dynamics within. As modeling progresses, disease by disease, it will be important to try to impose consistency in terms of how impacts are measured and to develop a uniform system along the lines of what others working in this area (Cutler, Rosen, and Vijan, 2006) are attempting to do.

Recommendation 6.1: A useful next step in the development of disease modeling for use in national health accounts would be to commission methodological research that develops a common language and frame of reference from which to start. A funding agency such as the National Institutes of Health or the National Science Foundation should consider supporting research to evaluate the merits and limitations of existing models for use in a variety of contexts, including national health accounting.

As part of this research, it will be important to remember that models require both a coherent theory of the natural history of the health condition and evidence regarding causal linkages among key variables. This points to the need for a strong body of comparative effectiveness research to inform the models. There is also a need to establish ongoing communication channels between the modelers, the data collection agencies, the comparative effectiveness researchers, and the policy makers who will ideally eventually be using the model results. Initial goals should include

- the development of a preliminary list of criteria against which alternative model integration options can be rated and compared,
- the development of a preliminary list of modeling needs that addresses contemporary and future planning issues and policy considerations, and
- the development of a preliminary action plan for moving toward a decision about model integration for the future.

It is also important to compare macrodata and micro-level disease models to ensure that the latter are constrained by national totals, and there is more to learn in these sorts of macro-micro linkages. While national health accounts data are available only as aggregates, the microsimulation models desired are about individual behavior. One way to compare the models with the data is to compute the necessary aggregates, from direct simulation, Monte Carlo methods, etc. The problem is that many different models can give the same aggregated behavior, so this does not provide a powerful test to compare different models. At the same time, however, it is possible to work back (or down) from the aggregate data to individual behaviors. Such ecological inferences do not have unique solutions, but the aggregate data, if used intelligently, can usually put fairly tight constraints on the individual behaviors, and the microsimulation can then be directly checked against these constraints (Shalizi, 2004). Similar macro-micro linkages can be formed for mortality modeling purposes. However, it is less clear how to ensure that quality of life is adequately constrained to national totals.

6.4.3. Integrated Modeling Efforts Outside the United States

For policy purposes, the ultimate goal is to develop an integrated simulation and forecasting model that predicts multiple attributes of health and health care costs for a population based on risk factors, multiple diseases, uptake of medical therapies, and program interventions. In short, the goal is to model the health care sector. A number of such efforts have been initiated internationally that may serve as a guide.

As part of a recent systematic review of cardiovascular disease policy models, Capewell and colleagues consulted extensively with internationally recognized modeling experts, identifying several multidisease models in use or in development. The following model descriptions rely heavily on their report to the British Heart Foundation (Capewell et al., 2008). While this list is not complete, it does attempt to highlight all the high-quality, ongoing projects that might be used as examples of best modeling practice.

- Dynamo-Health Impact Assessment (Dynamo-HIA, the Netherlands). The purpose of this project is to design a stochastic, dynamic microsimulation with explicit risk-factor states for annual, population-based data that models varying chronic disease processes with a discrete time frame. The model is used to estimate health impact assessment—the effect of policy interventions by modifying key parameters in the model to make projections and to show the development of the disease process over time. The model aims to contribute to informed policy making, both on the European Union level and the national level, by providing an instrument that enables health experts to predict the magnitude of health consequences that result from changing health determinants. Data required by the instrument will generally be available for most countries. The focus of this model will be cardiovascular disease, diabetes, and cancer. Once developed, Dynamo-HIA will be publicly available via a website with instructions for use, including data sets on risk factors and diseases for an example application of the model. An example would be estimating potential health gains by reducing smoking, and then implementing a tax increase on tobacco. For more information, see http://www.dynamo-hia.eu/object_class/dyhia_aims.html.

- Archimedes (United States). The Archimedes model is a comprehensive, continuous, trial-validated simulation model of human physiology, diseases, behaviors, interventions, and health care systems originally developed at Kaiser Permanente and now marketed by the independent organization Archimedes, Inc. The model currently includes several cardiovascular diseases (e.g., coronary artery disease, congestive heart failure, stroke, hypertension, dyslipidemia) as well as many other conditions (e.g., cancer, asthma, obesity) in a single integrated system, enabling it to model comorbidities, syndromes, medications with multiple effects, and more. It can be used to explore the effects of a wide variety of health care interventions on the progression, logistics, and economic outcomes of major diseases in a complex health care system. Potential applications include the design of guidelines, analysis of best practices, estimation of return on investment of care management programs, setting of clinical targets, priority setting, strategic goals, forecasting, design of performance measures, and research design. For more information, see http://archimedesmodel.com/index.html.
- Program in Occupational Health and Environmental Medicine (POHEM, Canada). POHEM is a continuous time, longitudinal, microsimulation model of health and disease. Using equations and submodels developed at Statistics Canada as well as drawn from the medical literature, the model simulates representative populations and allows the rational comparison of competing health intervention alternatives in a framework that captures the effects of disease interactions. The model simulates representative populations and allows for a rational comparison of competing health intervention alternatives. POHEM has been building and incrementally adding disease-specific microsimulation models to a macro model of pensions and the life cycle. The first microsimulation model was of acute myocardial infarction. The purpose of the model is to project the disease burden and to compare the relative impact of various risk-factor modifications through lifestyle and drug therapies on outcomes and costs. Since the original model, several additional disease models (including colorectal cancer, osteoarthritis, and others) have been developed, and each has been added in a modular fashion to the macro model. The model also includes a validated prediction model of diabetes incidence; work on a broader diabetes model is ongoing with plans to implement it in POHEM. For more information, see http://www.statcan.ca/english/spsd/Pohem.htm.

- Rijksinstituut voor Volksgezondheid en Milieu (RIVM, National Institute of Health and the Environment of the Netherlands, Chronic Disease Model (CDM). The RIVM CDM is a multistate transition model based on life-table methods. It has been used to describe the cardiovascular morbidity and mortality effects due to changes in cardiovascular risk factors versus changes in the rate of uptake of cardiovascular therapies. The model includes major risk factors (cholesterol, systolic blood pressure, smoking, physical activity level, and body mass index) and relevant disease states (acute myocardial infarction, other coronary heart disease, stroke, and chronic heart failure). It has also been expanded to model 20 other diseases but does not include treatments. Related transition between states is possible due to changes between classes for any risk factor, incidence, remission and progress for the considered diseases, and mortality. The model describes the life course of cohorts in terms of changes between risk-factor classes and changes between disease states over the simulation time period. The main model outcome variables are incidence, prevalence, and mortality numbers specified by disease, and quality of life adjusted outcomes. For more information, see http://rivm.openrepository.com/rivm/.
- Prevent (the Netherlands). Prevent is a dynamic population model that can handle multiple risk factors and diseases simultaneously. The first version was developed by Louise Gunning-Schepers and Jan Barendregt in 1989. It linked risk-factor exposure to disease-specific mortality, and through that to total mortality. Changes in risk-factor exposure translate (by way of potential impact fractions) into changes in excess risk with time lags. Prevent calculates two scenarios, called "reference" and "intervention." These scenarios differ only in a risk-factor intervention the user can specify; consequently all the differences between the two scenarios can be attributed to that intervention (counterfactual). The model has been through a number of important upgrades, including the addition of disease-specific and total morbidity (as well as disability and health care costs). The current version (3.01) includes an important conceptual change—the difference between risk factors and diseases has largely become semantic. Risk factors can be risk factors for other risk factors, diseases can be risk factors for other diseases and for risk factors. The current version allows defining "causal web" risk-factor/disease relations (although does not encourage spaghetti like risk-factor specifications) but also simple relations, such as letting the incidence of stroke increase as a consequence of a higher prevalence of ischemic heart disease. A major extension is that risk factors can now be either categorical or continuous. Categorical risk factors can have as many categories as desired, each associated with a relative risk. Continuous risk factors can be of various distributions (normal, lognormal, Weibull) and have risk functions asso-

ciated with them. The risk function relates risk to level of exposure and can be either linear, two-piecewise linear, or logit. In addition to being calculated, population projections can now also be provided as an input. When the option of calculated projections is chosen, the reference and intervention population projections will differ as a consequence of different risk-factor prevalence. When the population projection is input, the two scenarios will use the same projection. For more information, see Barendregt (1999).

- World Health Organization (WHO) CHOosing Interventions that are Cost Effective (CHOICE). The WHO CHOICE program assembles regional databases on the costs, impact on population health, and cost effectiveness of key health interventions. It also provides a contextualization tool that makes it possible to adapt regional results to the country level. This work started in 1998 with the development of standard tools and methods that have been used to generate the regional databases. An extension of cost-effective analyses was developed to inform better policy decisions at the national or regional level, called "generalized cost-effectiveness analysis," which can be summarized in two propositions:
  (1) The costs and benefits of a set of related interventions should be evaluated with respect to a counterfactual. This provides sufficient information for evaluating both independent and mutually exclusive options to identify the health-maximizing combination of interventions for a given budget.
  (2) Results of the analysis should initially be presented in a single league table. Subsequently, the decision would be made about the appropriate cut point for classifying interventions as very cost effective, very cost ineffective, and somewhere in between.

  For more information, see http://www.who.int/choice/en/.
- HealthAgeingMOD (Australia). The HealthAgeingMOD model is a cost-benefit model system under development by an Australian Research Council grant. The system will comprise several chronic disease progression models (or modules) linked to an overarching umbrella microsimulation model representing the entire Australian population. The microsimulation model accounts for individual and family-level demographic, socioeconomic, and health risk factors; progression of health status; chronic diseases; health-related expenditures; and quality of life. It will provide 20-year projections of disease-specific incidences, prevalences, and progression with treatment costs of chronic diseases and comorbidities. The current model will account for cardiovascular disease and diabetes; others will be added once proven successful. The model will provide a more complete view of chronic disease/comorbidity costs and benefits in various scenarios. For more information, see http://www.acerh.edu.au/publications/ACERH_RR3.pdf; Walker, Butler, and Colagiuri (2008).

- IMPACT (United Kingdom). The original IMPACT model was a cell-based model for coronary heart disease with two modules. One was for primary prevention that allowed generating incident cases and modeled the effect of interventions aimed at upstream and downstream risk factors either at the population or the individual level. It could deal with trends in risk factors over time, age, and gender. The other module dealt with disease treatments, including all relevant patient groups (acute coronary syndromes, chronic angina, survivors of myocardial infarction and heart failure) and all the evidence-based treatments. Successive versions of IMPACT have helped to explain trends in coronary heart disease in diverse populations in well over a dozen countries (Unal et al., 2004; Ford et al., 2007), and projects continue in several other countries. Published outputs from the IMPACT model now include deaths postponed and life years gained, which are then attributed to treatments in individuals or to population risk factor reductions (principally smoking and, to a lesser extent, blood pressure and cholesterol; Unal, 2004). While useful for policy purposes, there is a tremendous policy need for information on quality of life and on cost-effectiveness.
- IMPACT2 (United Kingdom). Building on the original IMPACT model, the IMPACT2 model will simulate individual coronary heart disease patients using event-driven simulation software developed specifically for the model. A key feature of the project was consultation with policy makers at national, regional, and local levels; these consultations are guiding both model development and the web-based user interface design. The model has been designed to allow subsequent extension into other cardiovascular and chronic diseases. Like the original IMPACT model, IMPACT2 is being built with two modules: one for primary prevention and the other for management of those with coronary heart disease. By end of 2009, it is anticipated that a web-based version of IMPACT2 will be publicly available. For more information, see http://www.liv.ac.uk/ihia/index.htm.

6.5. TYPES OF HEALTH DATA AND STATISTICS AND CHALLENGES TO THEIR IMPROVEMENT

Although significant conceptual work remains to be done in the construction of a framework on which to build national health accounts, the single biggest challenge is to create the data infrastructure—through new data collection and, perhaps more importantly, coordination of existing sources—required to support the effort. The basic data components of such a project are built from a range of sources, including censuses, vital statistics, population-based surveys, and administrative data (such as insurance claims). All of these data are currently collected in some form in the United States; however, they differ greatly in their coverage,

data creation is essential and could help guide the tweaking of now fragmented surveys to make them more useful in the future.

The remainder of this chapter assesses (1) the data currently available that would be useful in the construction of national health accounts and for linking expenditures to health, (2) key data challenges to improving that data, (3) data gaps to be filled, and (4) options for how to proceed.

6.5.1. Expenditure Data

At this point, we have already discussed the attributes of NHEAs in detail. Here, we only point out that the scope of spending included in NHEAs typically differs from that embedded in the various health surveys conducted in the United States. For example, NHEAs includes total net revenues for all U.S. hospitals, as well as government tax appropriations, nonpatient operating revenues (such as from gift shops), and nonoperating revenues (such as interest income) (see https://www.cms.gov/NationalHealthExpendData/ for more information). MEPS and the Medicare Current Beneficiary Survey (MCBS), in contrast, are event driven; most of these expenditures would not get picked up in the surveys, as they are not associated with discrete patient utilization events. Furthermore, there is little or no information about the nature, extent, or impact of trade and tax subsidies involving health resources.

Currently, limited data are available for estimating expenditures on acute care. MCBS does have this for the Medicare population. Reconciliations with National Health Expenditures data indicate that MCBS-based estimates are low by around 15 percent. Nursing home surveys could be used to uncover a large part of the missing piece (for admission periods, not long term). Claims data could possibly be combined with nursing home survey to get duration. The Census Bureau's American Community Survey (ACS) could provide a sampling frame for the institutionalized population that could be adopted by health surveys.

6.5.2. Censuses and Vital Statistics

Population censuses provide information on fertility, mortality, and migration. These data elements are used in the denominator of mortality rate calculations and therefore are essential for population projections and forecasting. They also serve as the basis for life tables, which allow life-expectancy statistics to be calculated. Vital statistics systems are critical for population health monitoring, in that they provide continuous information on births and deaths by age, gender, race, and other dimensions. Birth registries, the characteristics of which vary by state system, track health indicators and contain the data on live births needed to calculate infant mortality rates.[8] Mortality registries provide information on gender, age, education, occupation, residence, and cause of death, facilitating monitoring of age-specific and age-standardized death rates for total and cause-specific mortality. Accurate death attribution is quite clearly essential since health policy requires information on what people are dying from that links end-of-life spending to an episode, allowing the calculation of specific rates stratified by such variables as gender, ethnicity, educational attainment, and place of residence.

Mortality statistics are regularly published by NCHS. In fact, the agency has a long history of tracking vital statistics, in conducting population-based surveys, such as the NHANES, and in tracking measures of nonfatal health such as quality of life.[9] However, until recently, there have been no systematic efforts to disaggregate the dimensions in a way that parallels what is done for national health spending. Indeed, the disease groupings in the nation's vital statistics systems (e.g., ICD-10) are not even the same as those in the national expenditure surveys (MEPS and MCBS use ICD-9-CM).

Overall mortality rates are believed to be very accurate, but cause-of-death data are less so. Because the agency has only one need for the data—it is used for their annual survey, "How Healthy Are Americans?"—it unlikely that they would initiate enhanced data collection. Still, accounting of cause of death needs to be improved, perhaps using autopsy samples or surveys that get at disease incidence before death. And, for its beneficiary population, the Medicare-denominator file information could be linked with MCBS to measure trends in mortality.

Standard death certificates request that a causal sequence of events leading to death be specified, beginning with the "immediate cause of death" (the final event leading to death), then the "intermediate causes of death" (those leading up to the immediate cause), and finally the "underlying cause of death" (the disease initiating the sequence of events). In a separate section, information is requested on "other significant conditions contributing to death but not resulting in the underlying cause" (this might include disease risk factors such as smoking or hyperlipidemia). Most physicians do not receive formal training in filing death certificates, and it is common for the underlying cause of death to be confused for the mechanisms of death. To give a sense of the scope of the problem, in a study at one major academic medical center, close to 60 percent of death cer-

[8]NCHS maintains a linked birth and infant death data set that includes information from the death certificates and from birth certificates for each infant under 1 year of age who dies in the United States, Puerto Rico, the Virgin Islands, and Guam. The purpose of the linkage is to use the many additional variables available from the birth certificate to conduct more detailed analyses of infant mortality patterns. The linked files, which permit detailed analyses of infant mortality patterns, include information from the birth certificate such as "age, race and Hispanic origin of the parents, birth weight, period of gestation, plurality, prenatal care usage, maternal education, live birth order, marital status, and maternal smoking linked to information from the death certificate such as age at death and underlying and multiple cause of death," see http://www.cdc.gov/nchs/linked.htm.

[9]Appendix describes many of the U.S. data sets that will play a key role in supplying data for a national health account.

tificates were completed incorrectly (Zumwalt and Ritter, 1987). In many cases, the mechanism of death (e.g., cardiac arrest) was listed without an underlying cause. Other problems related to the reliability and validity of cause-of-death data include fragmentary completion of death certificates (Kircher and Anderson, 1987; Zumwalt and Ritter, 1987), changes in nosology (Sorlie and Gold, 1987; Lindahl and Johnson, 1994), inaccuracy of listed diagnoses (Israel, Rosenberg, and Curtin, 1986), and variation in physicians' interpretation of the causal sequence contributing to death (Gittelsohn and Senning, 1979; Gittelsohn, 1982). The system for generating cause-of-death statistics could be improved by requiring physicians to prepare forms online using a program with proper prompts to sort through the information—this should be part of the electronic medical records initiative. There also needs to be a minimum standard U.S. certificate of death; states could choose to require additional information.

The known problems with cause-of-death data as currently collected in the United States are exacerbated when comorbidities are present. Furthermore, cause of death on death certificates is an inherently poor source of data for understanding the excess risk of death due to certain risk factors as many are frequently underreported. For example, among decedents with a known diagnosis of diabetes, diabetes is listed on the death certificate less than 40 percent of the time (Bild and Stevenson, 1992; McEwen et al., 2006). In contrast, some data suggest overcoding of certain conditions. One paper reports that physicians may use coronary heart disease as a default diagnosis when uncertain about the underlying cause of death, resulting in inflated estimates of mortality from this cause (Lloyd-Jones et al., 1998). In the Framingham Heart Study data, for example, coronary heart disease was overestimated as a cause of death 8-24 percent of the time overall and by as much as twofold in the elderly (Lloyd-Jones et al., 1998).

To improve the capacity to attribute death to specific diseases, researchers should explore the linkage between death data and national surveys. For a national health account, it is important to know what the impact on mortality is of specific events and conditions. In order for this to work, the project will likely need to pull data across years to have enough records to decompose deaths by condition.

Recommendation 6.2: The National Institutes of Health should commission research exploring true causes of death and compare the results against official death records and statistics.

Part of this research may involve linking National Death Index records to MEPS or NHANES to cross-check against disease records. Since researchers will need death statistics linked to data on other characteristics—such as week and month of death—this will require using the Census Data Centers for access as there is no public use option for these data. The Health Insurance Portability and Accountability Act (HIPAA) should not be a constraint to this work, since data should be available after the death of a respondent (although state laws trump in these cases).

6.5.3. National Surveys

We have already said a lot about population-sourced data (such as health interview surveys, epidemiological studies, and longitudinal studies), which generate information on factors such as health status, the presence of disabilities, and the use of medical care that is needed to monitor health outcomes and distributional equity.[10] Routine household surveys—such as MEPS and the National Health Interview Survey (NHIS)—and the individual surveys—such as MCBS—are frequently used to monitor disparities in health and in access to, and receipt of, health care. Furthermore, they are valuable in providing information about individuals from population groups that fall beyond typical data collection boundaries, such as children not yet in school, adults outside the labor force, individuals who are uninsured and, perhaps most importantly, individuals who have not accessed health care services.

Population Coverage

When considering the characteristics of a national health account, population coverage—and specifically the ability to include difficult-to-reach subpopulations and minorities—is a major issue. Although, taken together, the nation's health surveys provide extensive information on large portions of the U.S. population, no single instrument collects data on all subpopulations. Some at-risk populations are poorly captured in the nation's surveys, and efforts to identify or develop appropriate microdata for these populations should be a priority. Identifying all out-of-scope populations (and in-scope populations with insufficient data coverage) will allow a compendium of data needs to be created. Filling these needs could then be rank-ordered based on any combination of policy priorities—size of the population, size of the expenditures involved, health needs of the population, or policy priorities of the population, to name a few.

Recommendation 6.3: The key data-producing agencies (National Center for Health Statistics, Centers for Medicare & Medicaid Services, and Agency for Healthcare Research and Quality, with coordination from the Bureau of Economic Analysis) should work together to identify gaps in the data coverage of U.S. population subgroups and prioritize, fund, and promote research programs to fill them.

In terms of population coverage, the most notable (but not the only) omission from the combined MEPS-MCBS data is the non-Medicare institutionalized and active-duty military populations. These populations are not included in the NHIS, which serves as the sampling frame for other key surveys (MEPS, NHANES), thereby leaving out most medical costs for Alzheimer's patients and large costs for other diseases, particularly mental disorders. While the NHIS sampling frame

[10]Again, a wide range of such surveys are described in Appendix A.

(which MEPS uses) could be extended to include the institutionalized population, contracting with CMS to add the population to MCBS might be a more cost-effective option. They have a well-validated survey for this group as they already survey institutionalized individuals or their proxies.

It is the view of this panel that, in most cases, expanding current surveys is a better value in terms of data generated relative to creating new surveys. Either way, the goal is to have in place a nationally representative (except for uninsured) set of claims for the non-Medicare population. The Census Bureau's ACS sampling frame for institutionalized could possibly be exploited.

Recommendation 6.4: The Medical Expenditure Panel Survey (MEPS) should be updated to track as accurately as possible the use, spending, and outcomes for the non-Medicare portion of the population. If possible, data on a panel of patients should be maintained to minimize between-patient variation. A longitudinal structure would be desirable, but much can also be done with repeated cross-sections (the panel is refreshed annually).

For the military population, the U.S. Department of Defense (DoD) collects comprehensive data on the health and health care spending of all active-duty military personnel, although this information does not appear to be readily available to researchers on a centralized basis. Likewise, the U.S. Department of Veterans Affairs (VA) collects information for the 5 million veterans using VA health services. The VA and DoD have been tasked with developing a common electronic medical record but are only starting to consider its structure and contents. The two agencies should be brought into the health accounts data collection conversations; this should be a relatively easy step as both are already required to provide comparable data for their two agencies.

Better data are also needed to monitor progress toward eliminating disparities. Aside from census and vital statistics data, no other national data permit detailed study of racial or ethnic minorities outside the three largest subgroups (non-Hispanic whites, non-Hispanic blacks, and Hispanics). Because ethnic and racial subgroups often have very diverse health status and risk behaviors, it is essential that national and state health data capture trends for these subgroups.

Disease Prevalence

Another data deficiency inherent in the nation's surveys has to do with underestimation of disease prevalence and incidence.[11] A number of surveys (NHANES,

[11]Incidence measures the number of newly diagnosed cases during a specific time period. The incidence is distinct from prevalence, which refers to the number of cases in existence on a certain date. Incidence statistics require either knowledge that a disease is new or repeated measures and also a larger sample (if a disease lasts 10 years, then incidence is one-tenth of prevalence). It is a more current measure of prevention effects than is prevalence, but for many diseases it, too, depends on things that happened in the past.

MCBS, MEPS, and HRS for mental health) are useful for capturing health degradations associated with population disease incidence and risk factors. MCBS and MEPS already include some of the necessary longitudinal components—within each panel—for tracking these variables, but they are relatively short (MEPS is 2 years, MCBS is 3 years). HRS is also longitudinal. Even so, in addition to underestimation of the prevalence of many diseases, surveys tend to have trouble providing adequate information about low-prevalence diseases.

While national survey data, such as those generated from MEPS, are appropriate for high-prevalence illnesses, such as diabetes and cardiovascular disease, for low-prevalence conditions they tend to suffer from small sample sizes, especially for some demographic groups. For these low-prevalence conditions, additional data are required, often in the form of population- or disease-specific data sets and surveys. An example is the Surveillance, Epidemiology, and End Results Medicare data, which track expenditures for individuals with a range of cancer diagnoses. Another option increasingly explored by researchers is to combine the power of claims databases (convenience samples) with the representativeness of household surveys (probability samples), weighting the claims data to match the representativeness of the household surveys. The drawback to claims data, however, is that they are not readily accessible and are often expensive to obtain and edit to the level necessary for research.

For low-prevalence conditions, two specific types of data are needed in combination: (1) individual observations in a national probability sample (e.g., MEPS) that provide good measures for conditions that occur frequently enough in relatively small samples and for which a reliable statistical basis can be formed and (2) registries or other specialized clinical data for those rarer conditions that would otherwise require very large data sets to provide adequate informational detail. Another option is to combine the power of claims databases (convenience samples) with the representativeness of household surveys (probability samples), weighting the claims data to match the representativeness of the household survey. It would be a relatively low-cost endeavor to task CMS and AHRQ to pick one or two low-prevalence but high-priority conditions and pilot test these types of linkages. One hurdle here is that, for this research, one would ideally like to have access to a large claims database, such as the one maintained by Pharmetrics or Medstat, and this is likely to be difficult, or at least expensive.

Comparisons of MEPS, MCBS, and other national surveys (e.g., NHANES, NHIS, HRS, National Comorbidity Survey) indicate that prevalence rates are underestimated in the expenditures surveys for conditions that rely solely on utilization data for their identification (Rosen and Cutler, 2009). For example, individuals who suffer from a particular condition but do not have any attributed utilization events would not be captured by such a measure. Since the national surveys contain a few direct questions (i.e., Have you ever been told by a doctor you have ___?), it is known that the extent of underestimation varies from condition to condition. For example, among non-Medicare eligible adults in 2002, 4.6 percent responded positively for ever having been diagnosed with diabetes, and 4.3 percent had diabetes-related utilization. However, among the same population, 14 percent responded positively to having been diagnosed with arthritis, while only 0.63 percent had related utilization.

Because underestimation of disease prevalence is largely an expenditure survey problem only for conditions that rely solely on utilization data for identification, the problem can be readily rectified.

Recommendation 6.5: Once consensus has been reached on a disease classification system for use in the national health accounts and a minimum data set has been specified, existing population survey-based databases should add questions about these diseases in order to more accurately capture population-wide health and expenditure data for the core disease set.

6.5.4. State and Private-Sector Data Systems and Surveys

In addition to the federal health statistics surveys and programs, each of the 50 states and various private-sector entities maintain data systems and conduct surveys of hospitals, health providers, and health care organizations. The private sector includes organizations of health service providers, regional employer networks, health insurance plans, consumers, industry, and private philanthropy. These private organizations conduct many of the national and state data collection activities; quality across these sources is variable. These statistical efforts lead to duplication and overlap in the data systems in the public and private sectors—some of which is useful, and some of which is wasteful. Hospital inpatient data are collected in the public sector by CMS, NCHS, AHRQ, Substance Abuse & Mental Health Services Administration, VA, and others. At the same time, many states maintain their own hospital discharge data systems conducted by rate setting and planning offices and by health systems agencies. In the private sector, hospital data are collected by the American Hospital Association, many abstracting organizations, insurers, and health plans, among others. Although hospital data are important, their reporting burden is high; recording, storing, abstracting, and processing medical records are expensive for both the institution and the users. In many instances, the rationale for these overlapping and duplicative hospital inpatient data is difficult to justify (Rice, 2000).

Recommendation 6.6: The statistical agencies should explore the value and feasibility of acquiring private insurance data for use in the construction of medical care accounts. Major data gaps could be filled if private sources could be exploited.

For researchers and policy makers, there would be great value from pooled data from Aetna, Kaiser, and other third parties. These data sets provide large samples that allow for the collection of some information about spending on rare conditions. The major obstacle to getting the analog of Medicare claims for the privately covered population is that health insurance companies are unlikely to be willing to incur nontrivial information technology (IT) costs just to meet federal agency needs; however, if some of these expenses could be reimbursed, this problem might be surmounted. A further issue, which may not be resolvable, is that the insurers will not want to release transaction prices (it may be possible to get charges and utilization, but actual payments would be sensitive). The issues are similar to protecting rebate information in Medicare Part D data (in that case, there was statutory protection for release of the data). Finally, the companies probably would want some liability releases.

6.5.5. Administrative Data

One important purpose of this report is to begin identifying key data sources and to provide an indication of how these sources can be combined or better organized to add value to the national data infrastructure needed to move forward in a cost-effective manner. A potentially rich source of case-by-case data on expenditures and treatments is patient claims data. Medicare and Medicaid provide data coverage for large portions of the population. If claims forms included the "right" information, they would go a long way toward satisfying some health accounting data needs. A lot could be done with claims data for the insured population because of their enormous size and wide coverage.

In terms of linking outcomes to spending, data are already quite good for some diseases. Pharmaceutical companies collect data on the comparative effectiveness of new versus old technologies and new versus old drugs in order to make their case for reimbursement. Additional data are collected as part of clinical trials which, while limited in some senses, do provide evidence about quality changes occurring in medical care. In their influential heart attack study, Heidenreich and McClellam (2001) used clinical trial data to show that decreases in early mortality from myocardial infarction resulted from increased use of effective treatments. Data on outcomes will improve over time—ideally, BLS, BEA, and academic researchers would be able to take advantage of the millions of dollars spent in their collection. And some of these data and enhancements to the infrastructure could be developed within government sources.

Recommendation 6.7: Significant analytic value could be added to Medicare data compiled by the Centers for Medicare & Medicaid Services if more clinical information on outcomes and patient characteristics were included. It is possible that the addition of only a few clinical items to claim forms could greatly enhance performance measurement. The National Committee on Vital and Health Statistics, an advisory body to the secretary of health and human services, could provide guidance on this. The linking of tests and procedures to diagnoses would also be useful for confronting the comorbidity

problem. Even though adding lines to claim forms involves major political hurdles, the panel recommends that it be done.

More than three decades ago, RAND developed an experimental claims form that included clinical information (Newhouse and the Insurance Experiment Group, 1993), and it proved effective in defining patient episodes. This should provide encouragement that the task is feasible. However, we caution that complications will have to be dealt with. For example, cases arise in which the outcomes do not become apparent until years later. Without careful randomized control trials or other causal analyses, even with longitudinal data, it is not always possible to cleanly attribute outcomes to any particular intervention.

Overall, a reasonable strategy for constructing a core data set to underlie national health and medical care accounts would be to identify a currently available survey, or combination of surveys, such as MEPS, to serve as the backbone of the data infrastructure, then to use claims information in a supplemental role wherever population or disease coverage gaps appear. Claims data from Medicare and Medicaid would cover large portions of the population. However, there are some groups for which claims data will not be available—most obvious are the uninsured, who do not submit claims—so their spending would have to be measured another way. Patients from some kinds of institutions are also unexplored (BEA has been investigating ways of obtaining spending data for these groups).

Research is needed about how to design claim forms to be more useful for research and national accounting and specifically about how to make it easier to track episodes and outcomes. Clearly, electronic medical recordkeeping (many items are currently not coded) is essential as is some degree of standardization of forms. One could also imagine specific refinements that could be made as well, such as asking physicians to enter quality of life measures onto claims forms. Likewise, it would be helpful for research purposes to add a few select items to Medicare claims forms—e.g., questions about obesity, smoking, hypertension, or cholesterol. These advances in information content must be weighed against the additional burden that is created for doctors and to health care systems. We return to these ideas in the next section.

6.5.6. Disease Registries

A range of data is routinely collected to provide information about specific health conditions; examples include disease surveillance systems for notifiable conditions, such as HIV, and disease registries and health services statistics, which provide information for monitoring health status and health outcomes. One advantage of surveillance and notification systems and disease registers over other types of health data collections is that these sources can provide direct measures of incidence for particular diseases in the population. A drawback is that they provide data only on individuals who receive health care. Furthermore, the records may be poor or incomplete, depending on their source.

6.6. ORGANIZING THE DATA INFRASTRUCTURE

It would be hard to overemphasize the difficulty of linking expenditures on medical care and other health inputs to health outcomes. Even though various projects are under way and making progress on this front, too few of the connections have been accurately quantified to make these results a focal point for a report on national health accounts. That said, there clearly needs to be a consistent set of rules to serve as a blueprint for development of the data infrastructure designed for the purpose of, among other things, informing research on the causal factors that affect the population's health.

During the past few decades, researchers have begun focusing much more on this challenge, but what has developed is a somewhat arbitrary rule for defining episodes of treatment that would serve as the basis for linking costs to outcomes (for the medical care portion of the story). Although collecting outcomes data that can be linked to spending is a big challenge, it is not impossible—cost modeling has been conducted attempting to attribute spending to treatments and ultimately to change in health status. The National Institute of Economic and Social Research (NIESR)/York has been working with the United Kingdom's National Health Service to collect information on outcomes that can be matched with expenditures. In conjunction with this, NIESR has published recommendations on key health accounting topics, such as measuring health system output and productivity, quality adjusting health sector output, and valuing attributes of health care.[12] Research has also been done to provide a clinical basis for linking treatment, effect, and outcomes; a prototype UK account uses survival rates from a hospital episodes statistics survey and a sample of waiting times—missing quality of life for the most part (Lakhani, 2005).

Looking well into the future, creating the ability to link expenditures and health outcomes has to be an objective for the United States as well. It is clear that more research is needed before recommendations about a specific approach can be sensibly advanced. The ideal health statistics system would entail a "population-based, person-specific" data system for rigorous testing of hypotheses about the determinants of health (Roos and Shapiro, 1995; Roos et al., 1995). This system should include systematic data on health and health-related activities that can summarize health at the population and individual levels and should have coherence or "adding up" constraints (Wolfson, 1991).

The desired data elements can be organized into several categories: sociodemographics; incidence and prevalence of a select group of clinical conditions;

---

[12]Developing new approaches to measuring National Health Service outputs and productivity: Final Report. See http://www.niesr.ac.uk/pubs/searchdetail.php?PublicationID=664.

use of health services; health behaviors; health outcomes; and other nonhealth and/or nonmedical determinants of health. An economic accounting structure is needed so that inputs to health (most immediately, medical care), and output (population health) are tabulated independently. Improvements in health—both quantity and quality of life—are the most critical variables on the output side of a health statistics system designed to quantify the impact of health care spending.

6.6.1. The Need for a Computerized National Data Infrastructure

The United States is a long way off from being able to satisfactorily model population health and health determinants in a systematic way, but a starting point is to begin keeping better track of risk factors, health indicators, and other related data. It is encouraging that progress has been made along some fronts, but serious data gaps remain. The remainder of this chapter discusses options and makes recommendations on how to address some of the current data deficiencies in order to move toward more integrated and comprehensive "health statistics needed for the 21st century" (Rice, 2000).

To develop a national data infrastructure, operational definitions must be developed, and core and optional data elements must be determined. Pieces of the data infrastructure will need to be created and pilot tested, in particular the linkages between databases. This will require a coordinated developmental effort, far beyond the current scale of activities. It will take several years, a research network, and a core commitment from leaders to make this happen. While building this infrastructure will be costly, failure to do so will be far more so—taking a substantial toll on the health (and wealth) of the population and the ability of policy makers to rein in wasteful spending.

The World Health Organization outlines four priorities for data collection in support of producing national health accounts (World Health Organization, 2005): (1) to use all suitable existing data, (2) to adjust existing data to make them more suitable, (3) to improve or enrich surveys and administrative records to increase their suitability, and (4) to identify and arrange for the collection or generation of data that remain "missing." To recognize the practical constraints to data acquisition, we add a fifth goal (which really should precede the first): (5) to develop a schema to prioritize the acquisition of data.

Currently, for the United States, partial snapshots relating to the bigger attribution picture can be created. Spending on medical care can be tracked by provider and payer (NHEAs); global health and certain physiological measures are available (e.g., from such surveys as NHIS and NHANES, and HRS is promising); in addition, records on two universal health outcomes, birth and death, are collected. However, at this point, it is virtually impossible to link these things—spending on medical care to individual outcomes to population health effects—to expenditure data (in part because so many factors, in addition to medical care, affect health).

The development of a national data infrastructure designed to organize and coordinate current data and frame future data needs is critical if a national health account is to be developed. The matter appears to be a high priority for the current administration, which committed about $19 billion of the 2009 recovery package to encourage doctors and hospitals to install and use electronic health records.[13] The fact that a number of private-sector technology and telecommunications companies have embarked on related projects adds to the prospect that the country may soon have an effective e-health records system. Not only will this modernization, designed to link patient health histories as well as treatment guidelines, be helpful for research and policy purposes, but also it could, if properly deployed, improve care and help curb costs by helping stem unnecessary tests, reducing errors, and coordinating treatments.

Perhaps the most critical (and most difficult) piece in the development of a national health IT infrastructure will be creating a level of interoperability that approaches what exists for most other 21st century industries (see Box 6-2).

While it is beyond the scope of this panel to make specific IT infrastructure recommendations, we point out that long-term comprehensive strategies and short-term incremental strategies need not be viewed as mutually exclusive. While current data may not be ideal, policy makers can learn a great deal simply by using—and developing better mechanisms to link—data sets that already exist. Thus, the policy choice is to determine how to most effectively allocate information systems resources to meet both short- and long-term goals.

6.6.2. Longitudinal Data

While current surveys and administrative systems contain extensive cross-sectional data, a shortage exists for large-scale longitudinal data to better track health status and health events over the life course of individuals in the population. The high expense and respondent burden associated with longitudinal surveys means that any increases in longitudinal data collection will be incremental. Other countries, having confronted this same issue, have addressed the lack of longitudinal data with dynamic microsimulation models, as described above (Wolfson, 1991, 1995; Spielauer, 2007). When microsimulation models are used, longitudinal data are still needed to set the framework, but in much smaller quantities.

Recommendation 6.8: A study should be commissioned by a funding agency (National Institutes of Health or National Science Foundation) to take an inventory of other countries' population health statistics systems, the role played by microsimulation modeling, the implications for longitudinal

[13]Currently, only about 17 percent of physicians in the United States use computerized patient records, even though some very large service providers, such as Kaiser Permanente and the Mayo Clinic, do so. See http://www.nytimes.com/2009/09/10/technology/10records.html.

BOX 6-2
Interoperability and the U.S. Health Care Nonsystem

The health care system in America is not a system. Rather, it is a disconnected collection of large and small medical businesses, health care professionals, treatment centers, hospitals, and all those who provide support for them. Most players have their own internal structure for gathering and sharing information, but nothing ties those isolated structures into an interoperable national system capable of making information easily shared and compared.

Interoperable systems are invisible but essential, and people have come to depend on them in everyday life. Interoperable systems allow one person to speak to another using cell phones with different cellular service. ATM cards are good at virtually all banks nationwide and most banks internationally; they allow people to buy groceries and pay for gym memberships—all of this is possible because of secure interoperable systems. These systems work because the telephone and banking sectors have developed methods and standards that allow participants in their systems to easily access and exchange information while the companies operate independently and compete vigorously.

If banks (which require standards of privacy and confidentiality) can use interoperable systems, health care should be able to as well. The benefits of putting such a system, which will be dispersed across many stakeholders, in place surely outweigh the costs (which include any change in the risk of confidentiality disclosure). The banking industry probably did several of these same cost/risk-benefit analyses 20 years ago, but the lasting impact of the network of point-of-service banking and portability of finance is now clear.

NOTE: For more on this topic, see Walker et al. (2005). The authors estimate that interoperability and health information exchange could lead to $77 billion in savings. In addition, it would provide clinicians with full information at the point of care; it would enhance portability of information for patients; and it would give researchers a centralized data source.

data collection, and the advantages and disadvantages of the different approaches.

The study should be oriented toward providing guidance on how existing surveys, such as NHANES or MEPS, could be modified to optimize their analytic value. The ability to make NHANES longitudinal seems to exist now; it is just not done very often.[14] Data are released biannually—and with a quite short lag. A longitudinal component would be extremely useful, but the cost would be

[14]An example of an exception is the Epidemiologic Follow-Up Study, a national longitudinal study designed to investigate the relationships among clinical, nutritional, and behavioral factors assessed in the first NHANES.

high unless it was scaled back to include only a few targeted survey questions. It would also be useful to add NHANES measures into MEPS or to add claims information to NHANES.

Recommendation 6.9: The National Health and Nutrition Examination Survey, which is already conducted on an ongoing basis, should contain a longitudinal component to optimize its value as an input into a health account and into cost-effectiveness research. Due to confidentiality concerns, the longitudinal component (or at least publicly accessible data products derived from it) will need to be limited to key variables that do not unduly increase disclosure risks.

A trade-off exists in terms of the timing of longitudinal data collection. For example, it may be better to collect a sample that is twice as large every other year instead of annually. These options would have to be explored during the design phase. It would also be important to be able to link the data to Medicare records, which could perhaps be done in data centers. Some of this could be funded by reducing the scope of NHIS—which would just be used as a sampling frame—and then expanding the other surveys.

6.6.3. Data Sharing and Data Linkage

In considering future prospects for improved health statistics to meet policy needs, it must be acknowledged that resources will not grow in parallel to the demand for data and medical services. Budgetary pressures require that current data collection and dissemination procedures be constantly assessed. These trends imply that it is time for the statistical agencies to make stronger efforts to coordinate data collection efforts across surveys and agencies—which may entail a cultural shift from within—to invest in the 21st century vision for health statistics. The ability to exploit multiple data sources will be a key ingredient to the success of a national health account. However, there are a number of barriers to linking provider data on expenditures with data on patients. For example, some modernizing of key data sources is needed to ease the task of creating a linkable system. If public and private data are to be linked, a common identifier will be needed. And trends in spending and in outcomes need to be measured at same metric.

Because surveys are expensive and burdensome, cost efficiency requires linking across sources whenever possible. NHEA needs to be linkable to individual-level data contained in other health data sources (NHIS, HNANES, MCBS, MEPS, etc.).[15] Ideally, one would like to be able to link data from these key

[15]America's Health Insurance Plans is now working to standardize electronic personal health records with the hope that claims and other patient records (diagnoses, procedures, medications) can be carried from one insurer to another (see http://www.insurancetech.com/showArticle.jhtml?articleID=201400212). In addition, they could be linked to the National Death Index in principle.

many federal statistical agencies, such as NCHS, make considerable efforts to ensure data quality, the quality and reliability of much private-sector data are unknown. Because survey results are subject to sampling, reporting, processing, and nonresponse errors; data cannot be fully interpreted unless these errors are reported. While federal statistical surveys routinely report standard errors, they are often unavailable in reports emanating from facility and manpower private-sector surveys. Improvement in the quality and reliability of health statistics reported in the private sector is urgently needed.

6.6.5. Characteristics for a Minimum Data Set

In moving forward on development of a national health account, it is essential to identify the ideal data needs to inform policy and then to specify a minimum data set, designed to collect information consistently across all relevant surveys to ensure broad representativeness. Key long-term goals include standardizing questions representing the same concepts across surveys, identifying and filling data gaps (such as populations underrepresented in or absent from national surveys), and adding questions to surveys to obtain comparable data across survey populations.

Recommendation 6.10: A research project should be commissioned by the National Institutes of Health to identify the minimum data set of variables needed to support the infrastructure of the ideal health statistics system.

Ideally, the integrated population data set would include some biological measures, in addition to general health measures. This idea is not new—HRS already compiles biological and genetic data, so a model and some data exist. It would certainly be worthwhile to explore the possibility of enhancing MEPS with biological information. The central idea is to have data on both expenditures and health outcomes in the same place and by the same categories. Table 6-1 provides an example of what that data element list might look like.

As has been pointed out, a subset of nonmedical determinants of health will need to be collected routinely on surveys, and the research recommended above should provide guidance for the routine updating of surveys to generate data on key nonmedical influences on health. Funding agencies (e.g., NIH, National Science Foundation [NSF], Agency for Healthcare Research and Quality, etc.) should strongly consider supporting work designed to identify a small set of simple, low-burden questions that could be appended to surveys to learn more about such factors as physical environment and individual monitors (of stress, sleep hours, exercise, etc.) that affect health. The project should also catalogue the portions of the population that are and are not covered by the various data sources.

It is unrealistic to think that all of the data elements needed for a health account could be collected and coordinated in a centralized repository imme-

TABLE 6-1 Potential Data Elements for a Minimum Data Set

Domain	Possible Variables
Population Health Measures	
Mortality	
Generic HRQoL[a]	Specific measures TBD
Individual Characteristics	
Demographics	Age, gender, race/ethnicity, social security number, marital status
Socioeconomic status	Income, education, marital status, employment/work conditions and hours, social support networks, health literacy, social environment, physical environment, personal health practices, health services
Individuals: Physical and Mental Measures	
Anthropomorphic measures	Height, weight, waist circumference, blood pressure, heart rate
Chemistries	Creatinine, glucose, direct low-density lipoprotein cholesterol, etc.
Functioning	ADL, IADL, mobility, cognitive limitations
Individuals: Behaviors	
Smoking	Smoking/amount
Alcohol abuse	Ethanol intake
Physical activity	Indicators of sedentary lifestyle
Adherence to medical therapies	Medicines prescribed/taken
Health Services (ideally differentiating between acute and chronic)	
Primary care—outpatient	Hospital-based clinic, Y/N?; diagnoses for the encounter (ICD); procedure codes (CPT); provider ID; dates of service (for inpatient and output); type and other characteristics[b]
Specialty care—outpatient	Hospital-based clinic, Y/N? type?
Hospital services	
Hospital services—provider	
Pharmaceuticals	Prescription drugs taken chronically (probably more important than those taken short term).[c]
Other (the current National Health Expenditure Accounts service categories, plus others)	
Core set of diseases linkable to utilization data	ICD diagnoses or top 20-30 diseases[d]
Long-term care	
Palliative (end of life) care	Hospice, Y/N, home/elsewhere
Residential care (if included in the domain of medical care)	Assisted living, retirement homes

continued

TABLE 6-1 Continued

Domain	Possible Variables
Identifiers and other medical claim form information	Unique patient identifiers, ICD-9 codes for admission diagnosis, unique provider ID, dates of service, codes for identifying inpatient procedures performed, Healthcare Common Procedure Coding System for outpatient (such as lab work), service charges (amount billed), allowed amount, and paid amount.[e]
Identifiers and other pharmaceutical claim form information	Unique patient ID, Unique pharmacist ID, National Drug Code and quantity of specific medicine, prescribing physician identifier, allowed amount, paid amount
Identifiers and other hospital discharge abstract information	Hospital identifier, admission source, discharge status, admission and discharge dates, length of stay, type of secondary insurance, patient demographics, ICD-9-CM diagnosis and CPT codes, assigned diagnostic related group and major diagnostic category, total charges
Other Determinants of Health	
Environment	Climate, air quality, access to clean water
Geographic hospital data	Wages, supplies, other hospital input prices, payer mix, disproportionate share status, hospital type (academic, community), rural, urban

[a]HRQoL indicates health-related quality of life.
[b]For example, a measure of how concentrated the medical care market is.
[c]If it is not feasible to include entries for each drug, the top 20-30 (by sales) could simply be listed.
[d]There are many ways to approach which diseases are most important but, again, one simple way is to rank by cost of illness.
[e]Paid amount will be the hardest to get for commercial claims, but *for many purposes it is the most important*.

diately. The infrastructure will have to grow incrementally, and therefore data collection will have to be prioritized. In addressing how to pick which data to collect from a potentially endless list of factors (health, social, environmental, financial) that may affect current and future health and health care spending, potential criteria for the selection of a set of key indicators include

- the importance of what is being measured in terms of its impact on health status and health expenditures, the policy relevance, and the susceptibility of the problem to intervention;
- the scientific soundness of the measure in terms of its validity, reliability, and evidence base; and
- the feasibility and cost of obtaining nationally comparable data for the measure.

It will be important to develop a consensus on and formalize a set of criteria by which the selection of key data for additional collection, standardization, etc., is prioritized. The development of a functional, policy-responsive, integrated national health data system is clearly a long-term proposition. That said, it is an achievable goal made all the more possible by the ongoing collaborative efforts of the national data agencies. It is the hope of the panel that our full set of recommendations provides a logical set of building blocks that—together but implemented separately—would contribute significantly to a more coherent and policy-responsive system of health statistics.

6.6.6. Privacy Protection and Confidentiality

The research environment is increasingly complex, both because of rising public concerns about the privacy of individuals' personal information and because of the complexity and lack of understanding of the legislative and regulatory frameworks guiding the conduct of research using such personal information. Since the passing of HIPAA in 1996, there has been considerable confusion over how the patient privacy rules affect quality improvement studies. From a legal standpoint, HIPAA rules do not apply equally to everyone. Public health authorities can access patient data without consent in their efforts to prevent disease outbreaks, and hospitals can use data to improve quality of care. The problem, however, lies in defining the boundaries of these activities. When is quality improvement clinical care, and when is it research? To cite one among countless examples, is a study of an intervention to estimate the reduction in rates of catheter-related bloodstream infection (Pronovost et al., 2006) simply research or is it part of care? The problem is that the line is not clearly drawn, and immediate concerns over privacy seem to supersede the longer term needs for research. At a 2003 NSF workshop on confidentiality research, Peter Madsen described this tension as the "Privacy Paradox":

> The rush to ensure complete levels of privacy in the research context paradoxically results in less social benefit, rather than in more. . . . [T]hrough the additional concept of utility, people will recognize that while they surely have the right to privacy, they may also come to the realization that they have a duty to share information, if the common good is to be furthered.

In addressing HIPAA challenges to research, a balance must be struck between the public's right to know and to pursue improved health care and the right of individuals and institutions to protect their privacy.

The Institute of Medicine recently concluded that the HIPAA privacy rule is not only inadequate in safeguarding patient privacy, but also significantly impedes secondary research (Institute of Medicine, 2009). Privacy and the safeguarding of personal information against unauthorized disclosure are fundamental individual goods in terms of respecting personal dignity and protecting patients from discrimination. Privacy also holds societal value because it encourages individuals to participate in socially desirable activities like research. At the same time, research is an equally compelling individual and societal good that can help address some of the nation's most pressing health problems. Indeed, research on the determinants of health can help guide national efforts to focus life-saving interventions where they can do the most good in terms of improving individual and public health (Gostin and Nass, 2009).

Although informed consent is meant as a safeguard, it can also be a barrier to valuable research. The Institute of Medicine has proposed new rules that would make health research exempt from HIPAA privacy rules and emphasize data security, transparency, and accountability, regardless of the funding source. The proposed rules would add consistency in regulatory oversight and ensure protection of participants. In addition, the new system would include two alternatives to consent: (1) a system with ethical oversight to protect data privacy and security and (2) a certification system that would allow researchers to link deidentified data sets. The panel supports these recommendations.

Bibliography

Advisory Commission to Study the Consumer Price Index. (1996). *Toward a More Accurate Measure to the Cost of Living: Final Report to the Senate Finance Committee from the Advisory Commission to Study the Consumer Price Index.* M.J. Boskin, E. Dulberger, R.J. Gordon, Z. Griliches, and D.W. Jorgenson. Washington, DC: U.S. Government Printing Office. Available: http://www.ssa.gov/history/reports/boskinrpt.html [accessed May 2010].

Aizcorbe, A., and Nestoriak, N. (2008). *The Importance of Pricing the Bundle of Treatments.* Bureau of Economic Analysis Working Paper. Available: http://www.bea.gov/papers/pdf/wp2008-04_bundle_treatments_paper.pdf [accessed September 2010].

Aizcorbe, A.M., Flamm, K., and Khurshid, A. (2002). *The Role of Semiconductor Inputs in IT Hardware Price Decline: Computers vs. Communications.* Paper provided by the Board of Governors of the Federal Reserve Systems Finance and Economics Discussion Series No. 2002-37. Available: http://www.federalreserve.gov/pubs/feds/2002/200237/200237pap.pdf [accessed May 2010].

Aizcorbe, A.M., Retus, B.A., and Smith S. (2008). Toward a health care satellite account. *Survey of Current Business,* May, 24-30.

Akobundu, E., Ju, J., Blatt, L., and Mullins, C.D. (2006) Cost-of-illness studies: A review of current methods. *PharmacoEconomics*, *24*, 869-890.

American Medical Association. (2008). *A Framework for Measuring Healthcare Efficiency and Value.* Prepared by the Physician Consortium for Performance Improvement® Work Group on Efficiency and Cost of Care. Available: http://www.ama-assn.org/ama1/pub/upload/mm/370/framewk_meas_efficiency.pdf [accessed May 2010].

Andresen, E.M., Rothenberg, B.M., and Kaplan, R.M. (1998). Performance of a self-administered mailed version of the Quality of Well-Being (QWB-SA) questionnaire among older adults. *Medical Care*, *36*(9), 1349-1360.

AQA Alliance. (2006). *AQA Principles of "Efficiency" Measures.* Available: http://www.aqaalliance.org/files/PrinciplesofEfficiencyMeasurementApril2006.doc [accessed May 2010].

Arbitman, D.B. (1986). A primer on patient classification systems and their relevance to ambulatory care. *Journal of Ambulatory Care Management*, *9*, 58-81.

Archibald, R.B. (1977). On the theory of industrial price measurement: Output price indexes. *Annals of Economic and Social Measurement*, *6*, 57-72.

Arias, E. (2007). U.S. life tables, 2004. *National Vital Statistics Reports*, *56*(9), 16. Available: http://www.cdc.gov/nchs/data/nvsr/nvsr56/nvsr56_09.pdf [accessed May 2010].

Atkinson, T. (2005). *Atkinson Review: Final report. Measurement of Government Output and Productivity for the National Accounts.* Hampshire, England: Palgrave-Macmillan.

Averill, R.F., Goldfield, N.I., Wynn, M.E., McGuire, T.E., Jmullin, R.L., Gregg, L.W., and Bender, J.A. (1993). Design of a prospective payment patient classification system for ambulatory care—Medicare and Medicaid managed care: Issues and evidence. *Health Care Financing Review,* September. Available: http://findarticles.com/p/articles/mi_m0795/is_n1_v15/ai_15268435/?tag=content;col1 [accessed September 2010].

Balk, B.M. (1998). *Industrial Price, Quantity and Productivity Indices.* Boston: Kluwer Academic.

Banks, J., Marmot, M., Oldfield, Z., and Smith, J.P. (2006). Disease and disadvantage in the United States and in England. *Journal of the American Medical Association*, *295*, 2037-2045.

Berk, A., Paringer, L., and Mushkin, S.J. (1978). The economic cost of illness, fiscal 1975. *Medical Care*, *16*, 785-790.

Berlowitz, D.R., Rosen, A.K., and Moskowitz, M.A. (1995). Ambulatory care case-mix measures. *Journal of General Internal Medicine*, *10*, 162-170.

Berndt, E.R., Busch, S.H., and Frank, R.G. (1998). *Price Indexes for Acute Phase Treatment of Depression.* NBER Working Paper W6799. Cambridge, MA: National Bureau of Economic Research.

Berndt, E.R., Cutler, D.M., Frank, R.G., Griliches, Z, Newhouse, J.P., and Triplett, J.E. (2000). Medical care prices and output. In A.J. Culyer and J.P. Newhouse (Eds.), *Handbook of Health Economics* (vol. 1A, pp. 120-180). North Holland, Amsterdam: Elsevier.

Bild, D.E., and Stevenson, J.M. (1991). Frequency of recording of diabetes on U.S. death certificates: Analysis of the 1986 National Mortality Followback Survey. *Journal of Clinical Epidemiology*, *45*(3), 275-281.

Black, C., McGrail, K., Fooks, C., Baranek, P., and Maslove, L. (2004). *Data, Data Everywhere ...: Improving Access to Population Health and Health Services Research Data in Canada.* The Canadian Policy Research Networks and the Centre for Health Services and Policy Research. Available: http://www.cihr-irsc.gc.ca/e/28082.html [accessed March 2009].

Bloom, B.S., Bruno, D.J., Maman, D., and Jayadevappa, R. (2001). Usefulness of cost of illness studies in health care decision making. *PharmacoEconomics*, *19*, 207-213.

Bolin, K., Jacobson, L., and Lindgren, B. (2001). The family as the health producer: When spouses are Nash-bargainers. *Journal of Health Economics*, *20*, 349-362.

Brazier, J.E., and Roberts, J. (2004). The estimation of a preference-based measure of health from the SF-12. *Medical Care*, *42*(9), 851-859.

Brazier, J.E., Roberts, J., and Deverill, M. (2002). The estimation of a preference-based measure of health from the SF-36. *Journal of Health Economics*, *21*(2), 271-292.

Bureau of Economic Analysis. (1994). *A Satellite Account for Research and Development.* Available: http://bea.gov/scb/account_articles/national/1194od/maintext.htm [accessed September 2010].

Burnette, K., and Mokyr, J. (1995). The standard of living through the ages. In J.L. Simon (Ed.), *The State of Humanity* (pp. 135-147). Malden, MA: Blackwell.

Cai, L., Hayward, M.D., Saito, Y., Lubitz, J., Hagedorn, A., and Crimmins, E. (2003). Estimation of multi-state life table functions and their variability from complex survey data using the SPACE Program. *Demographic Research*, *22*, Art. 6. Available: http://www.demographic-research.org/Volumes/Vol22/6/22-6.pdf [accessed September 2010].

Capewell, S., Allender, S., Critchley, J., Lloyd-Williams, F., O'Flaherty, M., Rayner, M., and Scarborough, P. (2008). *Modeling the UK Burden of Cardiovascular Disease to 2020: A Research Report for the Cardio & Vascular Coalition and the British Heart Foundation.* London, England: British Heart Foundation. Available: http://www.heartofmersey.org.uk/uploads/documents/oct_08/hom_1223379066_Modelling_PDF_for_website_7th_.pdf [accessed May 2010].

Casale, A.S., Paulus, R.A., Selna, M.J., Doll, M.C., and Bothe, A.E. (2007). ProvenCare: A provider-driven pay-for-performance program for acute episodic cardiac surgical care. *Annals of Surgery, 246*(4), 613-623.

Catlin, A., Cowan, C., Heffler, S., and Washington, B. (2007). National health spending in 2005: The slowdown continues. *Health Affairs, 26*(January/February), 142-153.

Catron, B., and Murphy, B. (1996). Hospital price inflation: What does the new PPI tell us? *Monthly Labor Review, 119*(7), 24-31. Available: http://www.bls.gov/opub/mlr/1996/07/art3full.pdf [accessed May 2010].

Caves, D.W., Christensen, L.R., and Diewert, W.E. (1982). Multilateral comparison of output, input and productivity using superlative index numbers. *Economic Journal, 92*, 73-86.

Center on the Evaluation of Value and Risk in Health. (2007). *The Cost-Effectiveness Analysis Registry.* Tufts-New England Medical Center. Available: https://research.tufts-nemc.org/cear/default.aspx [accessed May 2010].

Centers for Disease Control and Prevention. (2010). *NCHS Data Linkage Activities.* Available: http://www.cdc.gov/nchs/data_access/data_linkage_activities.htm [accessed May 2010].

Champion, J.A.I. (1993). (Ed.). *Epidemic Disease in London.* Centre for Metropolitan History Working Papers Series, No. 1. London, England: Centre for Metropolitan History, University of London.

Chernew, M.E., and Joseph P. Newhouse (2008). What does the RAND health insurance experiment tell us about the impact of patient cost sharing on health outcomes? *American Journal of Managed Care, 14*, 412-414.

Christensen, L.R., and Jorgenson, D.W. (1969). The measurement of U.S. real capital input, 1919-1967. *Review of Income and Wealth, 15*(4), 293-320.

Christian, M.S. (2010). *Human Capital Accounting in the United States: 1994 to 2006.* University of Wisconsin Center for Education Research. Available: https://www.bea.gov/about/pdf/Christian_humancapital_20091101.pdf [accessed September 2010].

Clabaugh, G., and Ward, M.M. (2008). Cost-of-illness studies in the United States: A systematic review of methodologies used for direct cost. *Value in Health, 11,* 13-21.

Cohen, J.W., and Krauss, N.A. (2003). Spending and service use among people with the 15 most costly medical conditions, 1997. *Health Affairs, 22*(2), 129-138.

Colecchia, A., and Schreyer, P. (2002). The contribution of information and communication technologies to economic growth in nine OECD countries. *OECD Economic Studies, 34*, 154-171.

Cooper, B.S., and Rice, D.P. (1976). The economic cost of illness revisited. *Social Security Bulletin, 39*, 21-36.

Coresh, J., Selvin, E., Stevens, L.A., Manzi, J., Kusek, J.W., Eggers, P., Van Lente, F., and Levey, A.S. (2007). Prevalence of chronic kidney disease in the United States. *Journal of the American Medical Association, 298,* 2038-2047.

Cutler, D.M. (2004). *Your Money or Your Life: Strong Medicine for America's Health Care System.* Cary, NC: Oxford University Press.

Cutler, D.M., and Berndt, E.R. (Eds.). (2001). *Medical Care Output and Productivity.* Cambridge, MA: National Bureau of Economic Research.

Cutler, D.M., and McClellan, M. (2001). Is technological change in medicine worth it? *Health Affairs, 20*(September/October), 11-29.

Cutler, D.M., McClellan, M., Newhouse, J.P., and Remler, D. (1998). Are medical prices declining? Evidence for heart attack treatments. *Quarterly Journal of Economics, 113*(4, November), 991-1024.

Cutler, D.M., McClellan, M., Newhouse, J.P., and Remler, D. (2001). Pricing heart attack treatments. In D.M. Cutler and E.R. Berndt (Eds.), *Medical Care Output and Productivity* (National Bureau of Economic Research Studies in Income and Wealth Series, vol. 62, pp. 305-362). Chicago, IL: University of Chicago Press.

Cutler, D.M., Rosen, A.B., and Vijan, S. (2006). The value of medical spending in the United States, 1960-2000. *New England Journal of Medicine, 355*(August), 920-927.

D'Agostino, R.B., Vasan, R.S., Pencina, M.J., Wolf, P.A., Cobain, M., Massaro, J.M., and Kennel, W.G. (2008). General cardiovascular risk profile for use in primary care. The Framingham heart study. *Circulation, 117,* 743-753. Available: http://circ.ahajournals.org/cgi/reprint/117/6/743 [accessed May 2010].

Danzon, P.M., and Furukawa, M.F. (2006). Prices and availability of biopharmaceuticals: An international comparison. *Health Affairs, 25*(5), 1353-1362.

Danzon, P.M., and Furukawa, M.F. (2008). International prices and availability of pharmaceuticals in 2005. *Health Affairs, 27*(1), 221-233.

Danzon, P.M., Wang, Y.R., and Wang, L. (2005). The impact of price regulation on the launch delay of new drugs. *Health Economics, 14,* 269-292. Available: http://hc.wharton.upenn.edu/danzon/PDF%20Files/Impact%20of%20Price%20Regulation%20-%20Health%20Economics%202005.pdf [accessed May 2010].

Davis, K. (2007). Paying for care episodes and care coordination. *New England Journal of Medicine, 356,* 1166-1168.

Dawson, D., Gravelle, H., O'Mahony, M., Street, A., Weale, M., Castelli, A., Jacobs, R., Kind, P., Loveridge, P., Martin, S., Stevens, P., and Stokes, L. (2005). *Developing New Approaches to Measuring NHS Outputs and Productivity, Final Report.* York, England: Centre for Health Economics.

DeVol, R., and Bedroussian, A. (2007). *An Unhealthy America: The Economic burden of Chronic Disease: Charting a New Course to Save Lives and Increase Productivity and Economic Growth.* Washington, DC: Milken Institute. Available: http://www.milkeninstitute.org/publications/publications.taf?function=detail&ID=38801018&cat=resrep [accessed May 2010].

Diewert, E., and Mizobuchi, H. (2009). *An Economic Approach to the Measurement of Productivity Growth Using Differences Instead of Ratios.* Discussion Paper 09-03, Department of Economics, University of British Columbia. Available: http://faculty.arts.ubc.ca/ediewert/dp0903.pdf [accessed May 2010].

Douglas, J. (2006). *Measurement of Public Sector Output and Productivity*. New Zealand Treasury Policy Perspectives Paper No. 06/09. Available: http://www.treasury.govt.nz/publications/research-policy/ppp/2006/06-09/ [accessed May 2010].

Druss, B.G., Marcus, S.C., Olfson, M., Tanielian, T., Elinson, L., and Pincus, H.A. (2001). Comparing the national economic burden of five chronic conditions. *Health Affairs*, *20*(6), 233-242.

Eggleston, K., Shah, N.D., Smith, S.A., Wagie, A.E., Williams, A.R., Grossman, J.H., Berndt, E.R., Long, K.H., Bannerjee, R., and Newhouse, J.P. (2009). The net value of health care for patients with type 2 diabetes, 1997-2005. *Annals of Internal Medicine, 151*(6), 386-393.

Elixhauser, A., Steiner, C., and Palmer, L. (2007). *Clinical Classifications Software (CCS), 2007.* Washington, DC: Agency for Healthcare Research and Quality. Available: http://www.hcup-us.ahrq.gov [accessed May 2010].

Ellis, R.P., Pope, G.C., Iezzoni, L.I., Ayanian, J.Z., Bates, D.W., and Ash, A. (1996). Diagnosis-based risk adjustment for Medicare capitation payments. *Health Care Financing Review, 17*(3), 101-128.

Employers Health Coalition. (2000). *The Hidden Competitive Edge: Employee Health and Productivity.* Newton, MA: Managed Care Communications.

Englander, F., Hodgson, T.J., Terragrossa, R.A. (1996). Economic dimensions of slip and fall injuries. *Journal of Forensic Science, 41,* 733-746.

Erickson, P. (1998). Evaluation of a population-based measure of quality of life: The health and activity limitation index (HALex). *Quality of Life Research, 7*(2), 101-114.

Erickson, P., Kendall, E.A., Anderson, J.P., and Kaplan, R.M. (1989). Using composite health status measures to assess the nation's health. *Medical Care, 27*(3 Suppl.), S66-S76.

Erickson, P., Wilson, R., and Shannon, I. (1995). Years of healthy life. *Healthy People 2000 Statistical Notes, 7,* 1-15.

Etches, V., Frank, J., Di Ruggiero, E., and Manuel, D. (2006). Measuring population health: A review of indicators. *Annual Review of Public Health, 27,* 29-55.

Ettaro, L., Songer, T.J., Zhang, P., and Engelgau, M.M. (2004). Cost-of-illness studies in diabetes. *PharmacoEconomics, 22*(3), 149-164.

Eurostat. (2001). *Handbook on Price and Volume Measures in National Accounts.* Luxembourg, Germany: European Communities.

Evans, R.G., and Stoddart, G.L. (1990). Producing health, consuming health care. *Social Science and Medicine, 31,* 1347-1363.

Feenstra, R.C., and Shapiro, M.D. (2003) High-frequency substitution and the measurement of price indexes. In *Scanner Data and Price Indexes* (pp. 123-150). Cambridge, MA: National Bureau of Economic Research.

Feenstra, R.C., Diewert, E.W., and U.S. Office of Prices and Living Conditions (2001). *Imputation and Price Indexes: Theory and Evidence from the International Price Program.* Washington, DC: Bureau of Labor Statistics.

Feeny, D., Torrance, G., and Furlong, W. (1996). *Health Utilities Index: Quality of Life and Pharmacoeconomics in Clinical Trials.* Philadelphia, PA: Lippincott-Raven Press.

Feeny, D., Furlong, W., Torrance, G.W., Goldsmith, C.H., Zhu, Z., DePauw, S., Denton, M., and Boyle, M. (2002). Multiattribute and single-attribute utility functions for the health utilities index mark 3 system. *Medical Care, 40*(2) 113-128.

Fisher, E.S., Wennberg, D.E., Stukel, T.A., Gottlieb, D.J., Lucas, F.L., and Pinder, E.L. (2003). The implications of regional variations in Medicare spending: Health outcomes and satisfaction with care. *Annals of Internal Medicine, 138*(February 18), 288-298.

Fisher, F.M. (1993). *Aggregation: Aggregate Production Function and Related Topics.* Cambridge, MA: MIT Press.

Fisher, F., and Shell, K. (1972). *The Economic Theory of Price Indices: Two Essays on the Effects of Taste, Quality, and Technological Change.* New York: Academic Press.

Fixler, D.J. (1996). *The Treatment and Price of Health Insurance in the CPI.* Washington, DC: Bureau of Labor Statistics.

Fixler, D., and Ginsburg, M. (2001). Health care output and prices in the producer price index. In D.M. Cutler and E.R. Berndt (Eds.), *Medical Care Output and Productivity* (pp. 221-227). Cambridge, MA: National Bureau of Economic Research.

Fogel, R.W. (1986). Nutrition and the decline in mortality since 1700: Some preliminary findings. In S.L. Engerman and R.E. Gallman (Eds.), *Long-Term Factors in American Economic Growth* (pp. 439-555). Chicago: University of Chicago Press

Fogel, R. (2000). *Academic Economics and the Triumph of the Welfare State.* Address to the American Association of Universities, April 17, Washington, DC.

Ford, E.S., Ajani, U.A., Croft, J.B., Critchley, J.A., Labarthe, D.R., Kottke, T.E., Giles, W.H., and Capewell, S. (2007). Explaining the decrease in U.S. deaths from coronary disease, 1980–2000. *New England Journal of Medicine, 356,* 2388-2398.

Frank, R.G., Berndt, E.R., and Busch, S.H. (1999). Price indexes for the treatment of depression. In J.E. Triplett (Ed.), *Measuring the Prices of Medical Treatments* (pp. 72-102). Washington, DC: Brookings Institution Press.

Fraumeni, B., and Okubo, S. (2005). R&D in the national income and product accounts: A first look at its effect on GDP. In C. Corrado, J. Haltiwanger, and D. Sichel (Eds.), *Measuring Capital in the New Economy* (pp. 275-316). Chicago, IL: University of Chicago Press.

Friedman, D.J. (2006). *Assessing the Potential of National Strategies for Electronic Health Records for Population Health Monitoring and Research: Data Evaluation and Methods Research.* Vital Health Statistics, Series 2, No. 143, DHHS Pub. (PHS) 2007-1343. Hyattsville, MD: U.S. Department of Health and Human Services.

Friedman, M., and Kuznets, S.S. (1945). *Income from Independent Professional Practice.* New York: National Bureau of Economic Research.

Fryback, D.G., Dasbach, E.J., Klein, R., Klein, B.E., Dorn, N., Peterson, K., and Martin, P.A. (1993). The Beaver Dam health outcomes study: Initial catalog of health-state quality factors. *Medical Decision Making, 13*(2), 89-102.

Fryback, D.G., Dunham, N.C., Palta, M., Hanmer, J., Buechner, J., Cherepanov, D., Herrington, S.A., Hays, R.D., Kaplan, R.M., Ganiats, T.G., Feeny, D., and Kind, P. (2007). U.S. norms for six generic health-related quality-of-life indexes from the national health measurement study. *Medical Care, 45*(12), 1162-1170.

Fuchs, V.R. (1999), Health care for the elderly: How much? Who will pay for it? *Health Affairs, 18*(1), 11-22.

Garber, A.M., and Phelps, C.E. (1997). Economic foundations of cost-effectiveness analysis. *Journal of Health Economics, 16,* 1-31.

Garner, T.I., Janini, G., Passero, W., Paszkiewicz, L., and Vendemia, M. (2006). The CE and the PCE: A comparison. *Monthly Labor Review,* September, 20-46.

Gilbert, M. (1961). The problem of quality changes and index numbers. *Monthly Labor Review, 84*(9), 992-997.

Gittelsohn, A.M. (1982). On the distribution of underlying causes of death. *American Journal of Public Health, 72,* 133-140.

Gittelsohn, A.M., and Senning, J. (1979). Studies on the reliability of vital and health records: I. Comparison of cause of death and hospital record diagnoses. *American Journal of Public Health, 69,* 680-689.

Goetzel, R.Z., Long, S.R., Ozminkowski, R.J., Hawkins, K., Wang, S., and Lynch, W. (2004). Health, absence, disability, and presenteeism cost estimates of certain physical and mental health conditions affecting U.S. employers. *Journal of Occupational and Environmental Medicine, 46*(4), 398-412.

Gold, M.R. (1996). *Cost-Effectiveness in Health and Medicine*. New York: Oxford University Press.

Gold, M.R., Patrick, D.L., Torrance, G.W., Fryback, D.G., Hadorn, D.C., Kamlet, M.S., Daniels, N., and Weinstein, M.C. (1996). Identifying and valuing outcomes. In M.R. Gold, J.E. Siegel, L.B. Russell, and M.C. Weinstein (Eds.), *Cost-Effectiveness in Health and Medicine* (pp. 82-134). New York: Oxford University Press.

Gold, M.R., Stevenson, D., and Fryback, D.G. (2002). HALYS AND QALYS AND DALYS, OH MY: Similarities and differences in summary measures of population health. *Annual Review of Public Health, 23,* 115-134.

Goldenberg, K.L., Butani, S., and Phipps, P.A. (1993). Response-analysis surveys for assessing response errors in establishment surveys. In *American Statistical Association, Proceedings of the International Conference on Establishment Surveys,* June, Buffalo, NY, pp. 290-299. Alexandria, VA: American Statistical Association.

Gostin, L.O., and Nass, S. (2009). Reforming the HIPAA privacy rule: Safeguarding privacy and promoting research. *Journal of the American Medical Association, 301*(13), 1373-1375.

Graham, J.D., Thompson, K.M., Goldie, S.J., Segui-Gomez, M., and Weinstein, M.C. (1997). The cost-effectiveness of airbags by seating position. *Journal of the American Medical Association, 278,* 1418-1425.

Grazier, K.L. (2006). Efficiency/value-based measures for services, defined populations, acute episodes, and chronic Conditions. In Institute of Medicine, *Pathways to Quality Health Care, Performance Measurement: Accelerating Improvement* (Appendix H, pp. 222-249). Committee on Redesigning Health Insurance Performance Measures, Payment, and Performance Improvement Programs/Board on Health Care Services. Washington, DC: The National Academies Press. Available: http://books.nap.edu/openbook.php?record_id=11517&page=222 [accessed May 2010].

Grazier, K.L., Thomas, J.W., and Ward, K.A. (2002). *A Comparative Evaluation of Risk-Adjustment Methodologies for Profiling Physician Practice Efficiency*. A report to the Robert Wood Johnson Foundation. Ann Arbor, MI: Department of Health Management and Policy, University of Michigan.

Griliches, Z. (1957). Hybrid corn: An exploration in the economics of technological change. *Econometrica, 25,* 501-522.

Griliches, Z. (1964). Notes on the measurement of price and quality changes. In *Models of Income Determination* (National Bureau of Economic Research Studies in Income and Wealth, vol. 28, pp. 381-418). Princeton: Princeton University Press. Available: http://www.nber.org/chapters/c1823.pdf [accessed May 2010].

Griliches, Z. (1971). Hedonic price indexes of automobiles: An econometric analysis of quality change. In Z. Griliches (Ed.), *Price Indexes and Quality Change: New Methods of Measurement.* Cambridge, MA: Cambridge University Press.

Griliches, Z. (1992). (Ed.). *Output Measurement in the Service Sectors.* National Bureau of Economic Research Studies in Income and Wealth, vol. 56. Chicago: University of Chicago Press.

Grimm, B.T., Moulton, B.R., and Wasshausen, D.B. (2005). Information-processing equipment and software in the national accounts. In C. Corrado, J. Haltiwanger, and D. Sichel (Eds.), *Measuring Capital in the New Economy* (National Bureau of Economic Research Studies in Income and Wealth, pp. 363-402). Chicago: University of Chicago Press.

Grossman, M. (1972). On the concept of health capital and the demand for health. *Journal of Political Economy, 80*(2), 223-255.

Guterman, S., Davis, K., Schoenbaum, S., and Shih, A. (2009). Using Medicare payment policy to transform the health system: A framework for improving performance. *Health Affairs, 28*(2), w238-w250.

Hackbarth, G., Reischauer, R., and Mutti, A. (2008). Collective accountability for medical care—Toward bundled Medicare payments. *New England Journal of Medicine, 359*(1), 3-5.

Hanmer, J., Hays, R.D., and Fryback, D.G. (2007). Mode of administration is important in U.S. national estimates of health-related quality of life. *Medical Care, 45*(12), 1171-1179.

Hayward, R.A. (2008). Access to clinically-detailed patient information: A fundamental element for improving the efficiency and quality of healthcare. *Medical Care, 46*(3), 229-231.

Health Canada. (1998). *Economic Burden of Illness in Canada, 1998.* Policy Research Division, Strategic Policy Directorate, Population and Public Health Branch, Health Canada. Available: http://www.phac-aspc.gc.ca/publicat/ebic-femc98/pdf/ebic1998.pdf [accessed September 2010].

Heidenreich, P., and McClellan, M. (2001). Trends in treatment and outcomes for acute myocardial infarction: 1975-1995. *American Journal of Medicine, 110,* 165-174.

Heijink, R., Koopmanschap, M.A., and Polder, J.J. (2006). *International Comparison of Cost of Illness.* Bilthoven, The Netherlands: National Institute for Public Health and the Environment.

Heijink, R., Noethen, M., Renaud, T., Koopmanschap, M., and Polder, J. (2008). Cost of illness: An international comparison of Australia, Canada, France, Germany and The Netherlands. *Health Policy, 88*(1), 49-61.

Hicks, J. (1940). The valuation of social income. *Economica, 7,* 105-124.

Hodgson, T.A. (1988). Annual costs of illness versus lifetime costs of illness and implications of structural change. *Drug Information Journal, 22*, 323-341.

Hodgson, T.A., and Cohen, A.J. (1999). Medical expenditures for major diseases, 1995. *Health Care Financing Review, 21*(2), 119-164.

Hodgson, T.A., and Meiners, M. (1982). Cost-of-illness methodology; A guide to current practices and procedures. *Milbank Quarterly, 60*(3), 429-462.

Hopkins, J.R. (1999). Financial incentives for ambulatory care performance improvement. *Joint Commission Journal on Quality and Patient Safety, 25*(5), 223-238.

Hopkins, S. (2007). *OECD Handbook Measuring Education and Health Volume Output.* Draft Chapter 3 (on health outputs). Paper presented at the Workshop on Measuring Education and Health Volume, June 6-7, Paris, France. Available: http://www.oecd.org/dataoecd/1/40/38599465.pdf [accessed May 2010].

Hornbrook, M.C., Hurtado, A.V., and Johnson, R.E. (1985). Health care episodes: Definition, measurement, and use. *Medical Care Research and Review, 42*(2), 163-218.

Hornbrook, M.C., Goodman, M.J., Bennett, M.D., and Greenlick, M.R. (1991). Assessing health plan case mix in employed populations: Self-reported health status models. *Advances in Health Economics and Health Services Research, 12,* 233-272.

Hyman, D.A. (2009). *Health Care Fragmentation: We Get What We Pay For.* Illinois Law and Economics Research Paper LE09-012. Available: http://papers.ssrn.com/pape.tar?abstract_id=1377051 [accessed May 2010].

Iezzoni, L.I. (1997). Assessing quality using administrative data. *Annals of Internal Medicine, 127*(8), 666-674.

Iezzoni, L.I. (2003). (Ed.). *Risk Adjustment for Measuring Health Care Outcomes, Third Edition.* Chicago, IL: Health Administration Press.

Institute of Medicine. (1998). *Scientific Opportunities and Public Needs: Improving Priority-Setting and Public Input at the National Institutes of Health.* Committee on the NIH Research Priority-Setting Process Washington, DC: National Academy Press.

Institute of Medicine. (2006). *Valuing Health for Regulatory Cost-Effectiveness Analysis.* W. Miller, L.A. Robinson, and R.S. Lawrence (Eds.), Committee to Evaluate Measures of Health Benefits for Environmental, Health, and Safety Regulation. Board on Healthcare Services. Washington, DC: The National Academies Press.

Institute of Medicine. (2009). *Beyond the HIPAA Privacy Rule: Enhancing Privacy, Improving Health Through Research.* S.J. Nass, L.A. Levit, and L.O. Gostin, Eds. Committee on Health Research and the Privacy of Health Information: The HIPPA Privacy Rule. Board on Health Sciences Policy and Board on Health Care Services. Washington, DC: The National Academies Press.

Israel, R.A., Rosenberg, H.M., and Curtin, L.R. (1986). Analytical potential for multiple cause-of-death data. *American Journal of Epidemiology, 124,* 161-179.

Jones, C. (2001). The economic return to health expenditures. *FRBSF Economic Letter* 2001-36 (December 14). Federal Reserve Bank of San Francisco. Available: http://www.frbsf.org/publications/economics/letter/2001/el2001-36.html [accessed May 2010].

Jorgenson, D.W. (2001). Information technology and the U.S. economy. *The American Economic Review, 91*(1), 1-32.

Jorgenson, D.W., and Stiroh, K.J. (2000*). Raising the Speed Limit: U.S. Economic Growth in the Information Age.* Brookings Papers on Economic Activity 2000:1. Washington, DC: Brookings Institution Press.

Jorgenson, D.W., Gollop, F.M., and Fraumeni, B.M. (1987). *Productivity and U.S. Economic Growth.* Cambridge, MA: Harvard University Press.

Joumard, I., and Häkkinen, U. (2007). *Cross-Country Analysis of Efficiency in OECD Health Care Sectors: Options for Research.* OECD Economics Department Working Paper 554. Paris, France: Organisation for Economic Co-operation and Development. Available: http://www.olis.oecd.org/olis/2007doc.nsf/linkto/eco-wkp(2007)14 [accessed May 2010].

Kaplan, R.M., Sieber, W.J., and Ganiats, T.G. (1997). The quality of well-being scale: Comparison of the interviewer-administered version with a self-administered questionnaire. *Psychology & Health, 12*(6), 783-791.

Kazis, L.E., Miller, D.R., Clark, J.A., Skinner, K.M., Lee, A., Ren, X.S., Spiro, A., III, Rogers, W.H., and Ware, J.E., Jr. (2004a). Improving the response choices on the veterans SF-36 health survey role functioning scales: Results from the veterans health study. *Journal of Ambulatory Care Management, 27*(3), 263-280.

Kazis, L.E., Lee, A., Spiro, A., III, Rogers, W., Ren, X.S., Miller, D.R., Selim, A., Hamed, A., and Haffer, S.C. (2004b). Measurement comparisons of the medical outcomes study and veterans SF-36® health survey. *Health Care Financing Review, 25*(4), 43-58.

Kazis, L.E., Miller, D.R., Skinner, K.M., Lee, A., Ren, X.S., Clark, J.A., Rogers, W.H., Spiro, A., III, Selim, A., Linzer, M., Payne, S.M., Mansell, D., and Fincke, R.G. (2004c). Patient-reported measures of health: The veterans health study. *Journal of Ambulatory Care Management, 27*(1), 70-83.

Kindig, D.A, and Stoddart, G. (2003). What is population health? *American Journal of Public Health, 93*(3), 380-383.

Kindig, D.A., Asada, K., and Booske, B. (2008). Population health framework for setting national and state health goals. *Journal of the American Medical Association, 299*(17), 2081-2083.

Kircher, T., and Anderson, R.E. (1987). Cause of death: Proper completion of the death certificate. *Journal of the American Medical Association, 258,* 349-352.

Koopmanschap, M.A. (1998). Cost-of-Illness studies, Useful for health policy? *PharmacoEconomics, 14*(2), 143-148.

Koopmanschap, M.A., van Roijen, L., and Bonneux, L. (1991). *Cost of Illness in the Netherlands* (in Dutch). Rotterdam, The Netherlands: Erasmus University, Department of Public Health, Faculty of Medicine and Health Sciences.

Lakdawalla, D.N., Goldman, D.P., and Shang, B. (2005). The health and cost consequences of obesity among the future elderly. *Health Affairs, 24*(Suppl. 2), W5R30-W5R41. Available: http://content.healthaffairs.org/cgi/content/full/hlthaff.w5.r30/DC1 [accessed May 2010].

Lakhani, A., Coles, J., Eayres, D., Spence, C., and Rachet, B. (2005). Creative use of existing clinical and health outcomes data to assess NHS performance in England: Part I- performance indicators closely linked to clinical care. *British Medical Journal, 330,* 1426-1431.

LaPlante, M.P., Harrington, C., and Kang, T. (2002). Estimating paid and unpaid hours of personal assistance services in activities of daily living provided to adults living at home. *Health Services Research, 37*(2), 397-415.

Leapfrog Group for Patient Safety and Bridges to Excellence. (2004). *Measuring Provider Efficiency: Version 1.0.* White paper. Available: http://www.leapfroggroup.org/media/file/MeasuringProviderEfficiencyVersion1_12-31-2004.pdf [accessed May 2010].

Lee, D.W., Meyer, J.W., and Clouse, J. (2001). Implications of controlling for comorbid conditions in cost-of-illness estimates: A case study of osteoarthritis from a managed care system perspective. *Value Health, 4,* 329-334.

Light, D. (2008). *How reducing negligible risks drives up health costs: A closer look at the clinical trials for Crestor.* Posted to Science Progress, December 12, 2008. Available: http://www.scienceprogress.org/2008/12/how-reducing-negligible-risks-drives-up-health-costs/ [accessed November 2010].

Lindahl, B.I., and Johansson, L.A. (1994). Multiple cause-of-death data as a tool for detecting artificial trends in the underlying cause statistics: A methodological study. *Scandinavian Journal of Social Medicine, 22,* 145-158

Linder, F.E. (1966). The health of the American people. *Scientific American, 214*(6), 21-29.

Lloyd-Jones, D.M., Martin, D.O., Larson, M.G., and Levy, D. (1998). Accuracy of death certificates for coding coronary heart disease as the cause of death. *Annals of Internal Medicine, 129,* 1020-1026.

Loeppke, R., Taitel, M., Richling, D., Parry, T., Kessler, R.C., Hymel, P., and Konicki, D. (2007). Health and productivity as a business strategy. *Journal of Occupational and Environmental Medicine, 49*(7), 712-721.

Lu, K., and Tsiatis, A.A. (2005). Comparison between two partial likelihood approaches for the competing risks model with missing cause of failure. *Lifetime Data Analysis, 11,* 29-40.

Luft, H.S. (2006). *Overview of a Reconfigured Health System.* Paper presented to the Council on Health Care Economics and Policy at the Thirteenth Princeton Conference on Reinventing Health Care Delivery in the 21st Century, May 24-25, Robert Wood Johnson Foundation, Princeton, NJ.

MaCurdy, T., Kerwin, J., Gibbs, J., Lin, E., Cotterman, C., O'Brien-Strain, M., and Theobald, N. (2008). *Evaluating the Functionality of the Symmetry ETG and Medstat MEG Software in Forming Episodes of Care Using Medicare Data.* Burlingame, CA: Acumen. Available: http://www.cms.hhs.gov/Reports/downloads/MaCurdy.pdf [accessed September 2010].

Manning, W.G., Newhouse, J.P., Duan, N., Keeler, E.B., and Liebowitz, A. (1987). Health insurance and the demand for medical care: Evidence from a randomized experiment. *American Economic Review, 77*(3), 251-257.

Martin, S., Rice, N., and Smith, P.C. (2009). *The Link Between Healthcare Spending and Health Outcomes: Evidence from English Program Budgeting Data*. London, England: The Health Foundation.

Mathers, C., Vos, T., and Stevenson, C. (1999). *The Burden of Disease and Injury in Australia. Canberra: Australian Institute of Health and Welfare.* Available: http://www.aihw.gov.au/publications/index.cfm/title/5180 [accessed September 2010].

McClellan, M., and Newhouse, J. (1997). The marginal cost-effectiveness of medical technology: A panel instrumental-variables approach. *Journal of Econometrics, 77,* 39-64.

McDowell, I. (2006). *Measuring Health: A Guide to Rating Scales and Questionnaires.* New York: Oxford University Press.

McDowell, I., and Newell, C. (1996). *Measuring Health: A Guide to Rating Scales and Questionnaires.* New York: Oxford University Press.

McDowell, I., Spasoff, R.A., and Kristjansson, B. (2004). On the classification of population health measurements. *American Journal of Public Health, 94*(3), 388-393.

McEwen, L.N., Kim, C., Haan, M., Ghosh, D., Lantz, P.M., Mangione, C.M., Safford, M.M., Marrero, D., Thompson, T.J., Herman, W.H., and the TRIAD Study Group. (2006). Diabetes reporting as a cause of death: Results from the Translating Research Into Action for Diabetes (TRIAD) study. *Diabetes Care, 29*(2), 247-253. Available: http://care.diabetesjournals.org/content/29/2/247.full.pdf+html [accessed May 2010].

McGlynn, E.A. (2008). *Identifying, Categorizing, and Evaluating Health Care Efficiency Measures: Final Report.* Prepared by the Southern California Evidence-Based Practice Center, RAND Corporation, AHRQ Publication 08-0030. Rockville, MD: Agency for Healthcare Research and Quality.

McKeown, T. (1976). *The Role of Medicine: Dream, Mirage, or Nemesis?* London, England: Nuffield Provincial Hospitals Trust.

Meara, E., White, C., and Cutler, D. (2004). Trends in medical spending by age, 1963-2000. *Health Affairs, 23*(July/August), 176-183.

Mechanic, R.E., and Altman, S.H. (2009). Payment reform options: Episode payment is a good place to start. *Health Affairs*, *28*(2), 262-271.

Medicare Payment Advisory Commission. (2006). *Report to the Congress: Increasing the Value of Medicare.* Washington, DC: Author. Available: http://www.medpac.gov/documents/jun06_entirereport.pdf [accessed May 2010].

Medicare Payment Advisory Commission. (2008). A path to bundled payment around a hospitalization. In Medicare Payment Advisory Commission, *Report to the Congress: Reforming the Delivery System* (Chapter 4, pp. 83-103). Washington, DC: Author.

Meltzer, D. (2001). *Theoretical Foundations of Medical Cost-Effectiveness Analysis: Implications for the Measurement of Benefits and Costs of Medical Interventions.* Studies in Income and Wealth, Volume 62. Cambridge, MA: National Bureau of Economic Research.

Michaud, P-C., Goldman, D., Lakdawalla, D., Zheng, Y., and Gailey, A. (2009). *Understanding the Economic Consequences of Shifting Trends in Population Health.* NBER Working Paper 15231. Cambridge, MA: National Bureau of Economic Research. Available: http://ideas.repec.org/p/nbr/nberwo/15231.html [accessed May 2010].

Miller, H.D. (2007). *Creating Payment Systems to Accelerate Value-Driven Health Care: Issues and Options for Policy Reform.* New York: Commonwealth Fund. Available: http://www.commonwealthfund.org/Content/Publications/Fund-Reports/2007/Sep/Creating-Payment-Systems-to-Accelerate-Value-Driven-Health-Care--Issues-and-Options-for-Policy-Refor.aspx [accessed May 2010].

Miller, H.D. (2008). *From Concept to Reality: Implementing Fundamental Reforms in Health Care Payment Systems to Support Value-Driven Health Care, Version 2.0.* Issues for discussion and resolution at the 2008 Network for Regional Healthcare Improvement's Summit on Healthcare Payment Reform. Available: http://www.nrhi.org/downloads/2008NRHIPaymentReformSummitFramingPaper.pdf [accessed May 2010].

Miller, H.D. (2009a). *Better Ways to Pay for Healthcare: A Primer on Healthcare Payment Reform.* A report prepared for the 2008 Summit on Healthcare Payment Reform convened by the Network for Regional Healthcare Improvement and a part of the NRHI Healthcare Payment Reform Series. Available: http://www.rwjf.org/pr/product.jsp?id=37448 [accessed May 2010].

Miller, H.D. (2009b*). Episode Payment Systems: Their Structure and Potential for Transforming Health Care Cost and Quality.* Presentation to the Massachusetts Special Commission on the Health Care Payment System, February 24. Available: http://www.mass.gov/Eeohhs2/docs/dhcfp/pc/2009_02_24_Episode_Payment_Overview_Presentation.ppt [accessed May 2010].

Mokyr, J. (1997).Are we living in the middle of an industrial revolution? *Federal Reserve Bank of Kansas City Economic Review, 82*(2), 31-43.

Molla, M.T., Wagener, D.K., and Madans, J.H. (2001). Summary measures of population health: Methods for calculating healthy life expectancy. *Healthy People Statistical Notes, 21*(August), 1-11. Abstract available: http://www.ncbi.nlm.nih.gov/pubmed/11676467 [accessed May 2010].

Moyer, B. (2008). *National Accounting Issues.* Presentation at the 2008 Committee on National Statistics BEA Satellite Health Care Account Workshop, March 14, National Research Council, Washington, DC.

Moynihan, D.P. (1999). Data and dogma in public policy. *Journal of the American Statistical Association, 94*(June), 359-364.

Murphy, K.M., and Topel, R.H. (2006). The value of health and longevity. *Journal of Political Economy, 114*(October), 871-904.

Murray, C.J.L. (2002). *Summary Measures of Population Health: Concepts, Ethics, Measurement and Applications.* Geneva, Switzerland: World Health Organization.

Murray, C.J.L., and Lopez, A.D. (1996). Evidence-based health policy-lessons from the global burden of disease study. *Science, 274*(5288), 740-743.

Murray, R. (1992). Measuring public-sector output: The Swedish report. In Z. Griliches (Ed.), *Output Measurement in the Service Sector* (pp. 517-542). National Bureau of Economic Research Studies in Income and Wealth 56. Chicago, IL: University of Chicago Press.

National Bureau of Economic Analysis. (1980). *The Measurement of Capital,* D. Usher (Ed.), Conference proceedings, The Measurement of Capital, Toronto, Canada, 1976. National Bureau of Economic Research, Studies in Income and Wealth, vol. 45. Chicago, IL: University of Chicago Press.

National Center for Chronic Disease Prevention and Health Promotion. (2010). *Chronic Diseases are the Leading Causes of Death and Disability in the U.S.* (last updated December 2009). Atlanta, GA: Centers for Disease Control and Prevention. Available: http://www.cdc.gov/NCCdphp/overview.htm [accessed May 2010].

National Center for Health Statistics. (2007). *Health, United States, 2007, with Chartbook on Trends in the Health of Americans.* Hyattsville, MD: Author. Available: http://www.cdc.gov/nchs/data/hus/hus07.pdf [accessed May 2010].

National Research Council. (1979). *Measurement and Interpretation of Productivity.* Panel to Review Productivity Statistics. Committee on National Statistics, Assembly of Behavioral and Social Sciences. Washington, DC: National Academy of Sciences.

National Research Council. (1993). *The Epidemiological Transition: Policy and Planning Implications for Developing Countries.* J.N. Gribble and S.H. Preston, Eds. Committee on Population. Commission on Behavioral and Social Sciences and Education. Washington, DC: National Academy Press.

National Research Council. (2002). *At What Price? Conceptualizing and Measuring Cost-of-Living and Price Indexes*. Panel on Conceptual, Measurement, and Other Statistical Issues in Developing Cost-of-Living Indexes, C.L. Schultze and C. Mackie (Eds.), Committee on National Statistics, Division of Behavioral and Social Sciences and Education. Washington, DC: National Academy Press.

National Research Council. (2005). *Beyond the Market: Designing Nonmarket Accounts for the United States.* Panel to Study the Design of Nonmarket Accounts, K.G. Abraham and C. Mackie, Eds. Committee on National Statistics, Division of Behavioral and Social Sciences and Education. Washington, DC: The National Academies Press.

National Research Council. (2007). *Understanding Business Dynamics: An Integrated Data System for America's Future.* Panel on Measuring Business Formation, Dynamics, and Performance. J. Haltiwanger, L.M. Lynch, and C. Mackie, Eds. Committee on National Statistics. Division of Behavioral and Social Sciences and Education. Washington, DC: The National Academies Press.

National Research Council. (2009). *Strategies for a BEA Satellite Health Care Account: Summary of a Workshop*. C. Mackie, Rapporteur. Committee on National Statistics, Division of Behavioral and Social Sciences and Education. Washington, DC: The National Academies Press.

Network for Regional Healthcare Improvement. (2007). *Incentives for Excellence: Rebuilding the Healthcare Payment System from the Ground Up.* A summary report of NRHI's 2007 Summit on Creating Payment Systems to Accelerate Value-Driven Health Care, March 29, Pittsburgh, PA. Available http://www.nrhi.org/downloads/ROOTS_Incentives_for_Excellence.pdf [accessed May 2010].

Network for Regional Healthcare Improvement. (2008). *From Volume to Value: Transforming Health Care Payment and Delivery Systems to Improve Quality and Reduce Costs.* NRHI Healthcare Payment Reform Series. Pittsburgh, PA: Author. Available: http://www.rwjf.org/quality equality/product.jsp?id=36217&c=EMC-CA142 [accessed May 2010].

Neumann, P.J. (2005). *Using Cost-Effectiveness Analysis to Improve Health Care: Opportunities and Barriers.* New York: Oxford University Press.

Neumann, P.J., Rosen, A.B., and Weinstein, M.C. (2005). Medicare and cost-effectiveness analysis. *New England Journal of Medicine, 353,* 1516-1522.

Newhouse, J.P. (1992). Medical care costs: How much welfare loss? *Journal of Economic Perspectives*, *6*(3), 3-21.

Newhouse, J.P., and the Insurance Experiment Group. (1993). *Free for All? Lessons from the RAND Health Insurance Experiment*. Cambridge, MA: Harvard University Press.

Nordhaus, W.D. (2003). The health of nations: The contribution of improved health to living standards. In K. Murphy and R. Topel (Eds.), *Measuring the Gains from Medical Research: An Economic Approach* (pp. 9-40). Chicago, IL: University of Chicago Press.

Norlund, A., Apelqvist, J., Blitzen, P.O., Nyberg, P., and Schersten, B. (2001). Cost of illness of adult diabetes mellitus underestimated if comorbidity is not considered. *Journal of Internal Medicine*, *250,* 57-65.

Organisation for Economic Co-operation and Development. (2000). *A System of Health Accounts, Version 1.0.* Paris, France: Author. Available: http://www.oecd.org/dataoecd/41/4/1841456.pdf [accessed May 2010].

Organisation for Economic Co-operation and Development. (2001). *Measuring Capital: A Manual on the Measurement of Capital Stocks, the Consumption of Fixed Capital, and Capital Services.* Paris, France: Author.

Organisation for Economic Co-operation and Development. (2002). *OECD Manual: Measuring Productivity: Measurement of Aggregate and Industry-Level and Productivity Growth.* Paris, France: Author.

Organisation for Economic Co-operation and Development. (2009). *Input Document Unit 11. Estimating Expenditure by Disease, Age and Gender Under the System of Health Accounts (SHA) Framework.* SHA Document Code SHA-REV-11001 Paris, France: Author. Available: http://www.oecd.org/dataoecd/15/57/42098151.pdf [accessed May 2010].

Pacific Business Group on Health. (2005). *Advancing Physician Performance Measurement: Using Administrative Data to Assess Physician Quality and Efficiency.* Available: http://www.pbgh.org/programs/documents/PBGHP3Report_09-01-05final.pdf [accessed May 2010].

Patrick, D.L., and Erickson, P. (1993). *Health Status and Health Policy: Quality of Life in Health Care Evaluation and Resource Allocation.* New York: Oxford University Press.

Patrick, D.L., Bush, J.W., and Chen, M.M. (1973). Toward an operational definition of health. *Journal of Health & Social Behavior, 14,* 6-23.

Phelps, C.E. (1999). Comment. In J.E. Triplett (Ed.), *Measuring Prices of Medical Treatments* (pp. 108-117). Washington DC: Brookings Institution Press.

Philipson, T., and Posner, R. (2008). *Is the Obesity Epidemic a Public Health Problem?: A Decade of Research on the Economics of Obesity.* NBER Working Paper 14010. Cambridge, MA: National Bureau of Economic Research. Abstract available: http://www.nber.org/papers/w14010 [accessed May 2010].

Physician Consortium for Performance Improvement® Work Group on Efficiency and Cost of Care. (2008). *A Framework for Measuring Healthcare Efficiency and Value.* Available: http://www.ama-assn.org/ama1/pub/upload/mm/370/framewk_meas_efficiency.pdf [accessed May 2010].

Piwowar, H.A., Becich, M.J., Bilofsky, H., and Crowley, R.S, on behalf of the caBIG Data Sharing and Intellectual Capital Workspace. (2008). *Towards a Data Sharing Culture: Recommendations for Leadership from Academic Health Centers.* PLoS Medicine Policy Forum Open Access. Available: http://www.plosmedicine.org/article/info:doi/10.1371/journal.pmed.0050183 [accessed May 2010].

Poisal, J.A., Truffer, C., Smith, S., Sisko, A., Cowan, C., Keehan, S., Dickensheets, B., and the National Health Expenditure Accounts Projections Team. (2007). *Health Spending Projections Through 2016: Modest Changes Obscure Part D's Impact.* Available: http://content.healthaffairs.org/cgi/reprint/hlthaff.26.2.w242v1[accessed May 2010].

Polder, J.J., Meerding, W.J., Bonneux, L., and van der Maas, P.J. (2005). A cross-national perspective on cost of illness: A comparison of studies from the Netherlands, Australia, Canada, Germany, United Kingdom, and Sweden. *European Journal of Health Economics, 6*(3), 223-232.

Pronovost, P., Needham, D., Berenholtz, S., Sinopoli, D., Chu, H., Cosgrove, S., Sexton, B., Hyzy, R., Welsh, R., Roth, G., Bander, J., Kepros, J., and Goeschel, C. (2006). An intervention to decrease catheter-related bloodstream infections in the ICU. *New England Journal of Medicine, 355,* 2725-2732.

Rabin, R., and de Charro, F. (2001). EQ-5D: A measure of health status from the EuroQol Group. *Annals of Medicine, 33*(5), 337-343.

Rattray, M.C. (2010). *Measuring Healthcare Resources Using Episodes of Care.* Available: http://carevariance.com/images/Measuring_Healthcare_Resources.pdf [accessed May 2010].

Reinsdorf, M.B. (1993). The effect of outlet price differentials in the U.S. consumer price index. In M.F. Foss, M.E. Manser, and A.H. Young (Eds.), *Price Measurements and Their Use* (pp. 227-254). Chicago, IL: University of Chicago Press.

Reinsdorf, M.B., and Moulton, B.R. (1997). The construction of basic components of cost of living indexes. In T.F. Breshahan and R.J. Gordon (Eds.), *The Economics of New Goods* (pp. 297-436). National Bureau of Economic Research Studies in Income and Wealth, vol. 58, Chicago, IL: University of Chicago Press.

Rice, D.P. (1966). *Estimating the Cost of Illness.* Health Economics Series No. 6, Publication (PHS) 947-6. Rockville, MD: U.S. Department of Health, Education, and Welfare.

Rice, D.P. (2000). *Shaping a Vision for the 21st Century Health Statistics.* Presented at the National Committee on Vital and Health Statistics 50th Anniversary Symposium, June 20, Washington, DC. Available: http://www.ncvhs.hhs.gov/50thvision21stcent.htm [accessed May 2010].

Rice, D.P., and Horowitz, L.A. (1967). Trends in medical care prices. *Social Security Bulletin, 30*(July), 13-28.

Rice, D.P., and MacKensie and Associates. (1989). *Cost of Injury in the United States: A Report to Congress.* San Francisco, CA: Institute for Health and Aging, University of California and Injury Prevention Center, Johns Hopkins University.

Rice, D.P., Kelman, S., and Miller, L. (1991). Estimates of economic costs of alcohol and drug abuse and mental illness, 1985. *Public Health Reports, 106*(3), 280-292.

Ridker, P.M., Danielson, E., Fonseca, F.A., Genest, J., Gotto, A.M., Jr., Kastelein, J.J., Koenig, W., Libby, P., Lorenzatti, A.J., MacFadyen, J.G., Nordestgaard, B.G., Shepherd, J., Willerson, J.T., Glynn, R.J.; JUPITER Study Group. (2008). Rosuvastatin to prevent vascular events in men and women with elevated C-reactive protein. *New England Journal of Medicine, 359*(21):2195-2207.

Robert Wood Johnson Foundation. (2008). *Overcoming Obstacles to Health: Report from the Robert Wood Johnson Foundation to the Commission to Build a Healthier America.* Princeton, NJ: Author. Available: http://www.commissiononhealth.org/PDF/ObstaclesToHealth-Report.pdf [accessed May 2010].

Roehrig, C., Miller, G., Lake, C., and Bryant, J. (2009). National health spending by medical condition, 1996-2005. *Health Affairs, 28*(2), w358-w367.

Roos, N.P., and Shapiro, E. (1995). A productive experiment with administrative data. *Medical Care, 33*(12 Suppl.), DS7-DS12.

Roos, N.P., Black, C.D., Frohlich, N. Decoster, C., Cohen, M.M., Tataryn, D. Mustard, C.A., Toll, F., Carriere, K.C., Burchill, C.A, MacWilliam, L., and Bogdanovic, B. (1995) A population-based health information system. *Medical Care, 33*(12), 13-20.

Rosen, A.B., and Cutler, D.M. (2007). Measuring medical care productivity: A proposal for the U.S. National Health Account. *Survey of Current Business, 87,* 54-88.

Rosen, A.B., and Cutler, D.M. (2009). Challenges in building disease-based national health accounts. *Medical Care, 47*(7 Suppl. 1), S7-S13.

Rosen, A.K. (2001). Risk adjustment: Available tools and applications in primary care. In *Methods for Practice-Based Research Networks: Challenges and Opportunities* (pp. 137-168). Proceedings from the Practice-Based Research Networks Methods Conference convened by the American Academy of Family Physicians Task Force, Plan to Enhance Family Practice Research, November 29-December 1, San Antonio, TX. Available: www.aafp.org/online/etc/medialib/aafp_org/documents/clinical/research/pbrnmethods.Par.0001.File.tmp/pbrn_methods.pdf [accessed May 2010].

Rosen, A.K., and Mayer-Oakes, A. (1999). Episodes of care: Theoretical frameworks versus current operational realities. *Journal on Quality Improvement, 25*(3), 111-128.

Rosenberg, M.A., Fryback, D.G., and Lawrence, W.F. (1999). Computing population-based estimates of health-adjusted life expectancy: Comparison of alternative methods for computation. *Medical Decision Making, 19*(1), 90-97.

Rosenthal, M.B. (2008). Beyond pay for performance: Emerging models of provider-payment reform. *New England Journal of Medicine, 359*(12), 1197-1200.

Sandy, L.G., Rattray, M.C., and Thomas, J.W. (2008). Episode-based physician profiling: A guide to the perplexing. *Journal of General Internal Medicine, 23*(9), 1521-1524.

Sassi, F. (2006). Calculating QALYs, comparing QALY and DALY calculations. *Health Policy and Planning, 21*(5), 402-408.

Schenker, N., Gentleman, J.F., Rose, D., Hing, E., and Shimizu, I.M. (2002). Combining estimates from complementary surveys: A case study using prevalence estimates from national health surveys of households and nursing homes. *Public Health Reports, 117,* 393-407.

Schreyer, P. (2001). *OECD Manual on Productivity Measurement: A Guide to the Measurement of Industry-Level and Aggregate Productivity Growth.* Paris, France: Organisation for Economic Co-operation and Development.

Schreyer, P. (2008). *Output and Outcome: Measuring the Production of Non-market Services.* Paper presented at the 30th meeting of the International Association for Research in Income and Wealth, Portoroz, Slovenia.

Schreyer, P., Diewert, W.E., and Harrison, A. (2005). *Cost of Capital Services and the National Accounts.* Presented to the Meeting of the Canberra II Group on Non-Financial Assets in Canberra, March 29-April 1, Canberra, Australia.

Schulman, K.A., Yabroff, K.R., Kong, J., Gold, K.F., Rubenstein, L.E., Epstein, A.J., and Glick, H. (2001). A claims data approach to defining an episode of care. *Pharmacoepid and Drug Safety, 10,* 417-427.

Scitovsky, A.A. (1967). Changes in the costs of treatment of selected illnesses, 1951-1995. *American Economic Review, 57*(5), 1182-1195.

Selden, T.M., Levit, K.R., Cohen, J.W., Zuvekas, S.H., Moeller, J.F., Mckusick, D., and Arnett, R.H. (2001). Reconciling medical expenditure estimates from the MEPS and the NHA, 1996. *Health Care Financing Review, 23*(Fall), 161-178.

Sensenig, A.L., and Donahoe, G.F. (2006). Improved estimates of capital formation in the national health expenditure accounts. *Health Care Financing Review, 28*(1), 9-23.

Shalizi, C.R. (2004). Methods and techniques of complex systems science: An overview. In T.S. Deisboeck, J.Y. Kresh, and T.B. Kepler (Eds.), *Complex Systems Science in Biomedicine* (Chapter 1, pp. 33-114). New York: Springer. Available: http://arxiv.org/PS_cache/nlin/pdf/0307/0307015v4.pdf [accessed May 2010].

Shapiro, I., Shapiro, M.D., and Wilcox, D.W. (2001). Measuring the value of cataract surgery. In D.M. Cutler and E.R. Berndt (Eds.), *Medical Care Output and Productivity* (National Bureau of Economic Research Studies in Income and Wealth Series, vol. 62, pp. 411-438). Chicago, IL: University of Chicago Press.

Shaw, J.W., Johnson, J.A., and Coons, S.J. (2005). U.S. valuation of the EQ-5D health states: Development and testing of the D1 valuation model. *Medical Care, 43*(3), 203-220.

Shortell, S., and Casalino, L. (2008). Health care reform requires accountable care systems. *Journal of the American Medical Association, 300*(1), 95-97.

Shwartz, M., Ash, A., and Peköz, E. (2006). Risk adjustment and risk-adjusted provider profiles. *International Journal of Healthcare Technology and Management, 7*(1/2), 15-42.

Sing, M., Banthin, J., and Selden, T. (2006). Reconciling medical expenditure estimates from the Medical Expenditure Panel Survey and the national health accounts, 2002. *Health Care Financing Review, 28*(1), 25-40.

Siu, A.L., Spragens, L.H., Inouye, S.K., Morrison, S., and Leff, B. (2009). The ironic business case for chronic care in the acute care setting. *Health Affairs, 28*(1), 113-125.

Skinner, J.S., Fisher, E.S., and Wennberg, J. (2005). The efficiency of Medicare. In D.A. Wise (Ed.), *Analyses in the Economics of Aging* (pp. 129-160). National Bureau of Economic Research conference report. Chicago, IL: University of Chicago Press.

Skinner, J.S., Staiger, D.O., and Fisher, E.S. (2006). Is technological change in medicine always worth it? The case of acute myocardial infarction. *Health Affairs, 25*(2), w34-w47.

Slobbe, L.C.J., Heijink, R., and Polder, J.J. (2007). *Draft Guidelines for Estimating Expenditure by Disease, Age, and Gender Under the System of Health Accounts (SHA) Framework.* Bilthoven, The Netherlands: RIVM, Department for Public Health Forecasting, Erasmus MC, Department of Public Health, Faculty of Medicine and Health Sciences. Available: http://www.rivm.nl/vtv/object_binary/o6070_Draft%20Guidelines_Expenditure%20by%20disease,%20age%20and%20gender%20Dutch%20COI%20Study.pdf [accessed May 2010].

Smith, N.S., and Weiner, J.P. (1994). Applying population-based case-mix adjustment in managed care: The Johns Hopkins ambulatory care group system. *Managed Care Quarterly, 2,* 21-34.

Sorlie, P.D., and Gold, E.B. (1987). The effect of physician terminology preference on coronary heart disease mortality: An artifact uncovered by the 9th revision ICD. *American Journal of Public Health, 77,* 148-152.

Spielauer, M. (2007). Dynamic microsimulation of health care demand, health care finance, and the economic impact of health behaviors: Survey and review. *International Journal of Microsimulation, 1*(1), 35-53.

Stahl, J. (2008). Modelling methods for pharmacoeconomics and health technology assessment: An overview and guide. *PharmacoEconomics, 26*(2), 131-148.

Starfield, B., Shi, L., and Macinko, J. (2005). Contribution of primary care to health systems and health. *Milbank Quarterly, 83*(3), 457-502.

Statistics Canada. (2007). *National Population Health Survey: Health Institutions Component, Longitudinal. Data release, January 27.* Available: http://www.statcan.gc.ca/cgi-bin/imdb/p2SV.pl?Function=getSurvey&SDDS=5003&lang=en&db=imdb&adm=8&dis=2#a4 [accessed September 2010].

Stewart, A.L., and Ware, J.E.J. (1992). *Measuring Functioning and Well-Being: The Medical Outcomes Study Approach.* Durham, NC: Duke University Press.

Stewart, S.T., Woodward, R.M., Rosen, A.B., and Cutler, D.M. (2006). *A Proposed Method for Monitoring U.S. Population Health: Linking Symptoms, Impairments, and Health Ratings.* NBER Working Paper 11358. Cambridge, MA: National Bureau of Economic Research.

Stigler, G. (1961). *The Price Statistics of the Federal Government.* A report to the Office of Statistical Standards, Bureau of the Budget, General Series 73. New York: National Bureau of Economic Research.

Sullivan, D.F. (1971). A single index of mortality and morbidity. *HMSHA Health Reports, 86*(4), 347-355.

Thomas, J.W (2006). Should episode-based economic profiles be risk adjusted to account for differences in patients' health risks? *Health Services Research, 41*(2), 581-598.

Thomas, J.W., and Ward, K. (2006). Economic profiling of physician specialists: Use of outlier treatment and episode attribution rules. *Inquiry, 43*(3), 271-282.

Thomas, J.W., Grazier, K.L., and Ward, K. (2004). Comparing accuracy of risk-adjustment methodologies used in economic profiling of physicians. *Inquiry, 41,* 218-231.

Thorpe, K.E. (1999). Market incentives, plan choice, and price increases, *Health Affairs, 18*(6), 194-202.

Thorpe, K.E., Florence, C.S., and Joski, P. (2004). *Which Medical Conditions Account for the Rise in Health Care Spending?* Web exclusive (August 25). Available: http://content.healthaffairs.org/cgi/reprint/hlthaff.w4.437v1 [accessed May 2010].

Trajtenberg, M. (1989). The welfare analysis of product innovations, with an application to computed tomography scanners. *The Journal of Political Economy, 97*(2), 444-479.

Triplett, J.E. (1983). Concepts of quality in input and output price measures: A resolution of the user-value resource-cost debate. In M.F. Foss (Ed.), *The U.S. National Income and Product Accounts: Selected Topics*. National Bureau of Economic Research, Studies in Income and Wealth, vol. 47. Chicago, IL: University of Chicago Press.

Triplett, J.E. (1990). The theory of industrial and occupational classifications and related phenomena. In *1990 Annual Research Conference Proceedings* (pp. 9-25). Arlington, VA, March 18-21, Bureau of the Census. Washington, DC: U.S. Government Printing Office.

Triplett, J.E. (1997). Measuring consumption: The post-1973 slowdown and the research issues. *Federal Reserve Bank of St. Louis Review, 79*(3), 9-42.

Triplett, J.E. (2001). What's different about health?: Human repair and car repair in national accounts and national health accounts. In D.M. Cutler and E.R. Berndt (Eds.), *Medical Care Output and Productivity* (National Bureau of Economic Research Studies in Income and Wealth Series, vol. 62, pp. 15-96). Chicago, IL: University of Chicago Press.

Triplett, J.E. (2003). Using scanner data in the consumer price indexes: Some neglected conceptual considerations. In R.C. Feenstra and M.D. Shapiro (Eds.), *Scanner Data and Price Indexes* (National Bureau of Economic Research Studies in Income and Wealth Series, vol. 64, pp. 151-162). Chicago, IL: University of Chicago Press.

Triplett, J.E. (2011). Health systems productivity. In S. Glied and P. Smith (Eds.), *The Oxford Handbook of Health Economics*. New York: Oxford University Press.

Triplett, J.E., and Bosworth, B.P. (2004). *Productivity in the U.S. Services Sector: New Sources of Economic Growth*. Washington, DC: Brookings Institution Press.

Triplett, J.E., and Bosworth, B.P. (2008). The state of data for services sector productivity measurement. *International Productivity Monitor, 16*(Spring), 53-71.

Tynan, A., and Draper, D.A. (2008). *Getting What We Pay For: Innovations Lacking in Provider Payment Reform for Chronic Disease Care.* Research Brief 6, June. Washington, DC: Center for Studying Health System Change. Available: http://www.hschange.org/CONTENT/995/995.pdf [accessed May 2010]

Unal, B., Critchley, J.A., and Capewell, S. (2004). Explaining the decline in coronary heart disease mortality in England and Wales between 1981 and 2000. *Circulation, 109,* 1101-1107.

United Nations. (1993). *Handbook of National Accounting: Integrated Environmental and Economic Accounting.* Series F, No. 61. New York: U.N. Department for Economic and Social Information and Policy Analysis, Statistical Division.

Varmus, H. (2000). *Disease-Specific Estimates of Direct and Indirect Costs of Illness and NIH Support.* (November 1995, April 1997, February 2000). Washington, DC: National Institutes of Health.

Viscusi, W.K., and Aldy, J.E. (2003). *Labor Market Estimates of the Senior Discount for the Value of Statistical Life.* Washington, DC: Resources for the Future. Available: http://www.rff.org/documents/RFF-DP-06-12.pdf [accessed September 2010].

Walker, A.E., Butler, J., and Colagiuri, S. (2008). *Cost-Benefit Model System of Chronic Diseases in Australia to Assess and Rank Prevention and Treatment Options: Proposed Research.* ACERH Research Report 3. Canberra, Australia: Australian Centre for Economic Research on Health. Available: http://www.acerh.edu.au/publications/ACERH_RR3.pdf [accessed May 2010].

Walker, J., Pan, E., Johnston, D., Adler-Milstein, J., Bates, D.W., and Middleton, B. (2005). The value of health care information exchange and interoperability. *Health Affairs, 19*(January), 10-18.

Ware, J.E., Jr., Brook, R.H., Davies, A.R., and Lohr, K.N. (1981). Choosing measures of health status for individuals in general populations. *American Journal of Public Health, 71*(6), 620-625.

Weiner, J., Dobson, A., Maxwell, S., Coleman, K., Starfield, B., and Anderson, G. (1996). Risk-adjusted Medicare capitation rates using ambulatory and inpatient diagnoses. *Health Care Financing Review, 17,* 77-99.

Weinstein, M.C., O'Brien, B., Hornberger, J., Jackson, J., Johannesson, M., McCabe, C., and Luce, B.R. (2003). Principles of good practice for decision analytic modeling in health-care evaluation: Report of the ISPOR task force on good research practices—Modeling studies. *Value in Health, 6,* 9-17.

Williams, G., Baxter, R., Kelman, C., Rainsford, C., Hongxing, H., Gu, L., Vickers, D., and Hawkins, S. (2002). *Estimating Episodes of Care Using Linked Medical Claims Data.* Lecture Notes in Computer Science Series. Heidelberg, Germany: Springer Berlin. Available: http://www.springerlink.com/content/t6r5c8albb4uvydw/fulltext.pdf [accessed May 2010].

Wingert, T.D., Kralewski, J.E., Lindquist, T.E., and Knutson, D.J. (1995). Constructing episodes of care from encounter and claims data: Some methodological issues. *Inquiry, 32,* 162-170.

Winkelman, R., and Mehmud, S. (2007). *A Comparative Analysis of Claims-Based Tools for Health Risk Assessment.* Schaumburg, IL: Society of Actuaries.

Wolfson, M.C. (1991). A system of health statistics: Toward a new conceptual framework for integrating health data. *Review of Income and Wealth, 37*(1), 81-104.

Wolfson, M.C. (1995). *Socio-Economic Statistics and Public Policy: A New Role for Microsimulation Modeling.* Paper prepared for the 50th Session of the International Statistical Institute, August 21-29, Beijing, China. Statistics Canada Working Paper 81. Available: http://dsp-psd.pwgsc.gc.ca/Collection/CS11-0019-81E.pdf [accessed May 2010].

Woolf, S.H. (2009). Social policy as health policy. *Journal of the American Medical Association, 301*(11), 1166-1169.

World Cancer Research Fund and American Institute for Cancer Research. (2009). *Policy and Action for Cancer Research: Food, Nutrition, and Physical Activity, a Global Perspective.* London, England: World Cancer Research Fund.

World Health Organization. (2005). *A Guide to Producing National Health Accounts with Special Applications for Low-Income and Middle-Income Countries.* Geneva, Switzerland: Author. Available: http://www.who.int/nha/docs/English_PG.pdf [accessed May 2010].

World Health Organization. (2008). *Global Tuberculosis Control: Surveillance, Planning, and Financing.* Available: http://www.who.int/tb/publications/global_report/2008/summary/en/index.html [accessed May 2010].

World Health Organization. (2009). *International Shortlist for Hospital Morbidity Tabulation (ISHMT).* Available: http://www.who.int/classifications/icd/implementation/morbidity/ishmt/en/ [accessed May 2010].

World Health Organization Quality of Life Group. (1998a). Development of the World Health Organization WHOQOL-BREF quality of life assessment. *Psychological Medicine, 28*(3), 551-558.

World Health Organization Quality of Life Group. (1998b). The World Health Organization quality of life Assessment: Development and general psychometric properties. *Social Science and Medicine, 46*(12), 1569-1585.

Yuskavage, R.E. (1996). Improved estimates of gross product by industry, 1959-1994. Survey *of Current Business, 76*(8 August), 133-155.

Zhao, Y., Randall, P.E., Ash, A.S., Calabrese, D., Ayanian, J., Slaughter, J.P., Weyuker, L., and Bowen, B. (2001). Measuring population health risks using inpatient diagnoses and outpatient pharmacy data. *Health Services Research, 26*(6 Pt. 2), 180-193.

Zumwalt, R.E., and Ritter, M.R. (1987). Incorrect death certification: An invitation to obfuscation. *Postgraduate Medicine, 81*(8), 245-254.

Appendix A

The Potential Role of Various Data Sources in a National Health Account

This appendix is a partial list of data sources that could play roles in a national health account program as identified by panel members David Cutler and Allison Rosen.

The **Current Population Survey (CPS)** is a monthly sample survey of about 50,000 households conducted by the U.S. Census Bureau for the Bureau of Labor Statistics. The CPS (see http://www.census.gov/cps/) is the primary source of information on labor force characteristics of the U.S. population. Monthly estimates from the CPS include employment, unemployment, earnings, hours of work, and other indicators. The annual March supplement produces national and state estimates on health insurance coverage, including private health insurance, Medicare, Medicaid, and military health care (CHAMPUS).

The **Health and Retirement Study** is a longitudinal cohort study of health, retirement, and aging and is designed to assess changes in physical and mental functioning, family support resources, insurance coverage, financial well-being, labor market participation, and retirement planning in the elderly. The data provide mental health diagnoses, risk factors, and costs not available in other national data sources. Funded by the National Institute on Aging, this study administers surveys on more than 22,000 Americans over the age of 50 every 2 years (beginning in 1990). See http://hrsonline.isr.umich.edu/ for more information and public-use files.

The **Healthcare Cost and Utilization Project (HCUP),** conducted by the Agency for Healthcare Research and Quality (AHRQ), consists of several databases, including the State Inpatient Database (SID) and the Nationwide Inpatient Sample (NIS). SID contains data on all hospitals and all discharges from 39 participating states. AHRQ receives the data from each statewide data organization, processes them into a uniform format, and then returns the uniform SID files to the statewide data organization. Beginning in 1988 with inpatient data from 8 hospitals, the NIS database has grown to cover 38 states in 2006. The NIS comes with weights that can be used to produce national, regional, and state estimates for participating states. See http://www.ahrq.gov/data/hcup/datahcup.htm for the complete list of databases that constitute the HCUP.

The **Medical Expenditure Panel Survey (MEPS)** collects data similar to those of its predecessor, the National Medical Expenditures Survey. Beginning in 1996 and repeated on an annual basis, the MEPS samples a portion of the households that participate in the prior year's National Health Interview Survey. The household component of MEPS collects data continuously at both the person and household levels using an overlapping panel design. In 2002, approximately 18,000 households, consisting of nearly 40,000 individuals, were included in the survey. MEPS data are publicly available via download from the AHRQ website (see http://www.meps.ahrq.gov/mepsweb/).

Medicare claims are comprehensive medical claims records from 1991 through 2007 for about 50 million older Americans that are available to qualified entities from the Centers for Medicare & Medicaid Services (CMS). The files contain information on all Medicare-covered services (inpatient, hospital outpatient, ambulatory and physician care, durable medical equipment, home care, hospice care, skilled nursing care, rehabilitation care, etc.) and include information on diagnoses, treatments and procedures, discharge, Medicare and out-of-pocket expenditures, demographic characteristics, and managed care enrollment.

The **Medicare Current Beneficiary Survey (MCBS)**, sponsored by CMS, is a nationally representative survey of aged, disabled, and institutionalized Medicare beneficiaries. Panelists are followed over the span of 4 years and are interviewed three times each year, regardless of whether they live in a household or a long-term care facility or switch between the two during the course of the survey period. The MCBS began in 1991; it is a patient-level comprehensive source of information on the health status, health care utilization, and expenditures of fee-for-service Medicare beneficiaries, health insurance coverage, and socioeconomic and demographic characteristics of the entire spectrum of Medicare beneficiaries. Self-reported utilization and expenditure information undergoes extensive validation using Medicare claims data. The cost and use data are available through 2006; data on access to care are available through 2008 at http://www.cms.gov/mcbs/.

The **Medicare Denominator File**, maintained by the University of Minnesota's Research Data Assistance Center, contains demographic and enrollment information about each beneficiary enrolled in Medicare during a calendar year. Information contained in this file includes the beneficiary unique identifier, information on residence, date of birth, date of death, sex, race, and age. The files can be used to follow the vital status of the former respondents to the MCBS beyond their 3-year survey window. For the complete file's descriptions of data variables and terms of use, see http://www.resdac.umn.edu/ddde/dd_de.asp.

The **National Ambulatory Medical Care Survey** is designed to meet the need for objective, reliable information about the provision and use of ambulatory medical care services. Findings are based on a sample of visits to nonfederally employed office-based physicians who are primarily engaged in direct patient care. Physician subspecialists in anesthesiology, pathology, and radiology are excluded from the survey, which was fielded in 1973-1981, 1985, and 1989-present (most recent data are for 2007). Data are available at http://www.cdc.gov/nchs/ahcd.htm.

The **National Comorbidity Survey (NCS)/National Comorbidity Survey-Replication (NCS-R)** are nationally representative surveys of the mentally ill in the United States. These surveys use a structured diagnostic interview to measure the prevalence and correlates of mental disorders listed in the *Diagnostic and Statistical Manual of Mental Disorders, Third Edition, Revised*. The NCS was conducted in 1990-1992; the NCS-R was conducted in 2000-2002. The NCS and NCS-R make it possible to impute the prevalence of mental health disorders among the elderly and adult populations. See http://www.icpsr.umich.edu/cocoon/cpes/ncsr/sections/all/sections.xml for more information.

The **National Health Care Survey (NHCS)** is a family of National Center for Health Statistics (NCHS) provider-based surveys designed to meet the need for objective, reliable information about the organizations and providers that supply health care, the services rendered, and the patients they serve. The surveys are designed to answer key questions of interest to health care policy makers, public health professionals, and researchers. They may include the factors that influence the use of health care resources; the quality of health care, including safety; and disparities in health care services provided to population subgroups in the United States. Data from NHCS are organized around the settings in which health care is delivered. Data are accessible via the NCHS website (see http://www.cdc.gov/nchs/nhcs.htm).

The **National Health and Nutrition Examination Survey (NHANES)** is a nationally representative survey of the civilian noninstitutionalized population in the 50 states and the District of Columbia. The NHANES has been administered in four waves of varying duration since 1971. Beginning in 1999, updates to the NHANES have been released every 2 years. The survey contains a rich source of information on sociodemographic variables, preventive health behaviors, risk factors, clinical conditions, and health status indicators, including the results of lab tests, clinical examinations, and physical functioning tests. Downloads are available for data collected through the 2005-2006 cohort from the NCHS website (see http://www.cdc.gov/nchs/about/major/nhanes/datalink.htm).

National Health Expenditure Accounts are compiled annually by the Office of the Actuary at CMS. They provide total national spending across the health system by source of payment (public, private, and out-of-pocket) and service provider (hospitals, physicians, pharmaceutical firms, etc.). This information is summarized and published on a yearly basis in a "sources and uses" of funds table, available via download from the CMS website (see http://www.cms.hhs.gov/NationalHealthExpendData/) beginning with data from 1960.

The **National Health Interview Survey (NHIS)** is a cross-sectional, nationwide in-person survey of approximately 40,000 households representing the civilian noninstitutionalized U.S. population. The first survey was administered in 1957 and has been repeated on an annual basis. The NHIS allows the monitoring of trends in illness and disability and the tracking of progress toward achieving national health objectives. Annual updates of the NHIS are available via download from the NCHS website (see http://www.cdc.gov/nchs/nhis.htm).

The **National Health Measurement Study (NHMS)**, directed by the University of Wisconsin School of Medicine and Public Health, is a nationally representative sample of 3,844 noninstitutionalized adults ages 35-89. The NHMS was designed to compare several common instruments that attempt to assess an individual's health-related quality of life based on self-reported symptoms and impairments and related questions, which vary across instruments. Fielded in 2005-2006, the data set is available at http://www.disc.wisc.edu/NHMS/index.html.

The **National Home and Hospice Care Survey (NHHCS)** is a continuing series of surveys of home and hospice care agencies in the United States. Information was collected about agencies that provide home and hospice care and about their current patients and discharges. The NHHCS is based on a probability sample of home health agencies and hospices and includes all agencies licensed or certified by Medicare or Medicaid. The NHHCS was fielded in 1992-1994, 1996, 1998, 2000, and 2007. See http://www.cdc.gov/nchs/nhhcs.htm for more information and updates.

The **National Hospital Ambulatory Medical Care Survey** collects data on the utilization and provision of ambulatory care services in hospital emergency and outpatient departments. Findings are based on a national sample of visits to emergency departments and outpatient departments of noninstitutional general and short-stay hospitals, exclusive of federal, military, and Department of Veterans Affairs (VA) hospitals. The survey spans the 50 states and the District of Columbia and has been fielded every year since 1992 (the most recent year available is 2007). Data are available at http://www.cdc.gov/nchs/ahcd.htm.

The **National Hospital Discharge Survey**, conducted annually since 1965, provides nationally representative data on inpatients discharged from nonfederal short-stay hospitals in the United States. Only hospitals with an average length of stay of fewer than 30 days for all patients, general hospitals, and children's general hospitals are included in the survey. Federal, military, and VA hospitals, as well as hospital units of institutions (such as prison hospitals) and hospitals with fewer than six beds staffed for patient use are excluded. In 2005, approximately

375,000 inpatient records were obtained from 444 hospitals. The most recent data are for 2006, available at http://www.cdc.gov/nchs/nhds.htm.

The **National Household Survey on Drug Abuse (NHSDA)**, conducted by the Substance Abuse and Mental Health Services Administration, focuses on the incidence, prevalence, consequences, and patterns of substance use and abuse. In 1997, the NHSDA was expanded from 18,000 to about 25,000 respondents to generate estimates for the nation and for two states (California and Arizona). In 1999, the NHSDA was further expanded to 70,000 respondents to generate estimates for all 50 states. Summary data and annual reports through 2008 can be found at http://www.oas.samhsa.gov/nsduhLatest.htm.

The National Medical Expenditure Survey (NMES) was a year-long panel survey of approximately 15,000 households consisting of nearly 36,000 individuals in the civilian, noninstitutionalized population. The survey provides information on the U.S. population's access to health care, use of health services, expenditures and sources of payment for care, health insurance coverage, health status, risk factors and disease prevalence, demographic characteristics, and employment and economic status. The NMES is publicly available via download from the Inter-university Consortium for Political and Social Research (see http://www.icpsr.umich.edu/).

The **National Nursing Home Survey (NNHS)** is a national survey of nursing homes, their residents, and discharged patients conducted by the NCHS. Surveys were administered in 1973-1974, 1977, 1985, 1995, 1997, 1999, and 2004. The NNHS provides information on the institutionalized population residing in nursing homes regardless of the resident's insurance status and the primary payer of the nursing home stay. The NNHS data are available via download from the NCHS website at http://www.cdc.gov/nchs/nnhs.htm.

The **National Survey of Ambulatory Surgery (NSAS)** is the only national study of ambulatory surgical care in hospital-based and freestanding ambulatory surgery centers. The NSAS was conducted from 1994 to 1996 but discontinued due to lack of resources. The NSAS was again conducted in 2006, with results released in 2010 at http://www.cdc.gov/nchs/nsas.htm.

The **National Survey of Family Growth** is a periodic survey of women ages 15-44 that began in 1973—men were added to the survey in 2002. The survey gathers information on family life, marriage and divorce, pregnancy, infertility, use of contraception, and men's and women's health. The survey results are used by the U.S. Department of Health and Human Services and others to plan health services and health education programs and to determine the need for other statistical studies of families, fertility, and health. Data and latest information are available on the NCHS website (see http://www.cdc.gov/nchs/NSFG.htm).

The **National Vital Statistics System** collects and disseminates information on vital events, which include births, deaths, marriages, divorces, and fetal deaths. Registrars in 57 vital event registration areas collect data from local officials and transmit them to NCHS. NCHS compiles those data and issues public-use data files and analytical reports that have been used in health services research to estimate life expectancy and mortality rates for specific diseases. See http://www.cdc.gov/nchs/nvss.htm for more information.

The **Surveillance, Epidemiology, and End Results–Continuous Medicare History Sample File (SEER-CMHSF)** has annual Medicare spending for 5 percent of the Medicare population and includes data on medical expenses for inpatient hospital stays, outpatient services, skilled nursing facility stays, home health agency charges, and physician services. The linked SEER-CMHSF contains all Medicare claims for the cancer patients included in the SEER registries. These data are available from 1976 to 1997 and then 2001 and 2002. These data can be used to determine the cost and health impact of cancer care by service category. The SEER-CMHSF contains roughly 70,000 cancer patients per year and 400,000 noncancer control individuals per year.

The **Surveillance, Epidemiology, and End Results–Medicare Linked Database (SEER-Medicare)** reflects the linkage of two large population-based sources of data that provide detailed information about Medicare beneficiaries with cancer. Over a dozen state and regional registries across the United States participate in the SEER program by reporting all newly diagnosed cases of cancer and tracking these individuals until their deaths. Beginning in 1975, data from 9 registries are available, covering approximately 10 percent of the U.S. population. From 2000 onward, 17 registries report data covering 26 percent of the population. The information collected by the program includes detailed clinical data and the patients' demographic characteristics. The data provide information on incidence, costs, and health outcomes of cancer patients. The SEER-Medicare database provides a linkage between the SEER cancer registries and Medicare enrollment and claims files. See http://healthservices.cancer.gov/seermedicare/ for more information.

The **World Health Survey (WHS)** was administered in 2002-2003 to monitor whether health systems were achieving their desired goals. The WHS was the first major survey program designed to address the issue of cross-population comparability in measuring health. Samples were randomly selected (for ages older than 18 years) and number between 1,000 and 10,000 individuals in each of 70 participating countries. See http://www.who.int/healthinfo/survey/en/index.html for results and analysis.

Appendix B

Biographical Sketches of Panel Members and Staff

JOSEPH P. NEWHOUSE (*Chair*) is the John D. MacArthur professor of health policy and management and chair of the Committee on Higher Degrees in Health Policy in the Malcolm Wiener Center for Social Policy at the Kennedy School of Government of Harvard University. He is a member of the faculties of the John F. Kennedy School of Government, the Harvard Medical School, the Harvard School of Public Health, and the Faculty of Arts and Sciences, as well as a faculty research associate of the National Bureau of Economic Research. He spent the first 20 years of his career at RAND, where he designed and directed the RAND Health Insurance Experiment, a project that from 1971 to 1988 studied the consequences of different ways of financing medical services. From 1981 to 1985 he was head of the RAND economics department. His expertise is in the areas of health care financing, health research policy, health services research, health care quality and outcomes, and general economics/ health economics. He is a member of the Institute of Medicine. He has B.A. and Ph.D. degrees in economics from Harvard University.

DAVID M. CUTLER is the Otto Eckstein professor of applied economics in the Department of Economics and the Kennedy School of Government and was formerly an associate dean of the Faculty of Arts and Sciences for social sciences at Harvard University. He is also a research associate at the National Bureau of Economic Research. His research concentrates on the value of medical innovation. He has examined how population health is changing over time, the importance of medical and nonmedical factors in improved health, and the value of increased medical spending. He has written extensively arguing that medical care is more productive than current statistics indicate and that the medical care cost problem is overstated. He is also interested in the economics of health insurance and the impact of managed care on the medical system. Cutler served on the Council of Economic Advisers and the National Economic Council during the Clinton administration and advised the presidential campaigns of Bill Bradley and John Kerry. Among other affiliations, he has held positions with the National Institutes of Health and has served on many study groups of the National Research Council/National Academy of Sciences. He is a member of the Institute of Medicine. He has a Ph.D. in economics from the Massachusetts Institute of Technology.

DENNIS G. FRYBACK is professor emeritus of population health sciences and industrial engineering at the University of Wisconsin. He is a founding member of the Society for Medical Decision Making, has been continuously active in its work since 1978, served as president in 1982-1983, and received its EL Saenger Service Award in 1994 and Award for Career Achievement in 1999. His research and teaching interests include medical technology assessment, health care cost-effectiveness analysis, measurement of population-level health status and health-related quality of life assessment, use of simulation modeling to understand cancer epidemiology, and use of Bayesian statistical analysis in these areas and in pharmacoeconomics and outcomes research. He was initiator and program director for an institutional doctoral training grant in population-based health services research at the University of Wisconsin–Madison. In 1986 he succeeded the first editor-in-chief of *Medical Decision Making* for a 3-year term. He currently serves on the editorial board of *Health Services and Outcomes Research Methodology.* He is fellow of the Association of Health Services Research. He is a member of the Institute of Medicine. He has a Ph.D. in psychology from the University of Michigan.

ALAN M. GARBER is the Henry J. Kaiser, Jr., professor and professor of medicine at Stanford University and staff physician at the U.S. Department of Veterans Affairs, Palo Alto Health Care System. At Stanford, he is also professor of economics and of health care delivery and financing. He is the founding director of the Center for Health Policy and the Center for Primary Care and Outcomes Research. His research focuses on methods for improving health care delivery and financing—particularly for the elderly—in settings of limited resources. He has developed methods for determining the cost effectiveness of health interventions, and he studies ways to structure financial and organizational incentives to ensure that cost-effective care is delivered. In addition, his research explores how clinical practice patterns and health care market characteristics influence technology adoption, health expenditures, and health outcomes in the United States and other countries. He leads the Global Healthcare Productivity project, which includes collaborators from 19 nations. He is a member of the Institute of Medicine. He has A.B., M.S., and Ph.D. degrees, all in economics, from Harvard

University and an M.D. from the Stanford School of Medicine. He completed a residency in medicine at Brigham and Women's Hospital in Boston.

EMMETT B. KEELER is a professor in the Department of Health Policy at the Pardee RAND Graduate School and senior mathematician at RAND. He leads a large study to evaluate a new model for helping people with chronic diseases manage their health better. He also directs a project at the University of California, Los Angeles, that supplies cost-effectiveness analyses to a variety of geriatric interventions and a project to develop a business case for providers to offer higher quality care. In the RAND Health Insurance Experiment, he investigated the theoretical and empirical effects of alternative health insurance plans on episodes of treatment and health outcomes. He has led several projects dealing with the potential demand for and effects of medical savings accounts. He has received article-of-the-year awards from the Association for Health Services Research for papers on outlier payments (1988), on the costs to others of bad health habits (1989), and on whether impoverished Medicare patients receive worse care in hospitals than do other patients (1994). He is the author or coauthor of numerous articles and four books. He has a Ph.D. in mathematics from Harvard University.

CHRISTOPHER D. MACKIE (*Study Director*) is a staff officer with the Committee on National Statistics (CNSTAT) specializing in economic measurement and statistics. He has served as study director for a number of projects, including those that produced the following reports: *At What Price?: Conceptualizing and Measuring Cost-of-Living and Price Indexes* (2002), *Beyond the Market: Designing Nonmarket Accounts for the United States* (2005), and *Understanding Business Dynamics: An Integrated Data System for America's Future* (2007). He has also led a number of CNSTAT initiatives related to data access, sharing, and confidentiality issues. Previously he was a senior economist with SAG Corporation, where he conducted a variety of econometric studies in the areas of labor and personnel economics, primarily for federal agencies. He is the author of *Canonizing Economic Theory*. He has a Ph.D. in economics from the University of North Carolina and has held teaching positions at the University of North Carolina, North Carolina State University, and Tulane University.

ALLISON B. ROSEN is assistant professor in the Division of General Medicine and the Department of Health Management at the University of Michigan. She is also clinical director of the Center for Value-Based Insurance Design in the University of Michigan School of Public Health. Her clinical work is primarily in the areas of diabetes, geriatrics, and cardiovascular disease. Her research interests focus on the impact of drug benefit design on the quality and value of health care. She has an M.D. from Duke University, an M.P.H. from the University of North Carolina, Chapel Hill, and an Sc.D. from Harvard University. She completed a residency at the University of California, San Francisco, and is certified by the American Board of Internal Medicine. She has received fellowships in general medicine and health policy from Harvard Medical School, the Health Services Research unit of the Agency for Healthcare Research and Quality, and the Harvard School of Public Health.

JACK E. TRIPLETT is a visiting fellow at the Brookings Institution in Washington, DC. His current research concerns productivity in health, finance, and other services industries, with a focus on developing improved measures of output for these notably difficult to measure sectors of the economy. From 1985 to 1997, he was chief economist at the U.S. Bureau of Economic Analysis (on leave in 1996-1997 to the National Bureau of Economic Research). From 1971 to 1985, he held positions at the U.S. Bureau of Labor Statistics, including associate commissioner for research and evaluation and chief of the Price Research Division. In 1979, he was assistant director for price monitoring at the Council on Wage and Price Stability. He has written extensively on problems of economic measurement, including price indexes, national accounts, capital stock and labor input, and productivity and technical change. He is the editor of *Fifty Years of Economic Measurement* (with Ernst R. Berndt), *The Measurement of Labor Cost*, and *Measuring the Prices of Medical Treatments*. He is a fellow of the American Statistical Association and the 1997 winner of the Julius Shiskin Award for Economic Statistics, awarded jointly by the National Association of Business Economists and the Washington Statistical Society. He has A.B., M.A., and Ph.D. degrees from the University of California, Berkeley.

COMMITTEE ON NATIONAL STATISTICS

The Committee on National Statistics (CNSTAT) was established in 1972 at the National Academies to improve the statistical methods and information on which public policy decisions are based. The committee carries out studies, workshops, and other activities to foster better measures and fuller understanding of the economy, the environment, public health, crime education, immigration, poverty, welfare, and other public policy issues. It also evaluates ongoing statistical programs and tracks the statistical policy and coordinating activities of the federal government, serving a unique role at the intersection of statistics and public policy. The committee's work is supported by a consortium of federal agencies through a National Science Foundation grant.